The South, the North and the Environment

THE SOUTH, THE NORTH AND THE ENVIRONMENT

PETER CALVERT AND SUSAN CALVERT

PINTER
London and New York

PINTER
A Cassell Imprint
Wellington House, 125 Strand, London WC2R 0BB
370 Lexington Avenue, New York, NY 10017-6550

First published 1999

© Peter and Susan Calvert 1999

British Library Cataloguing in Publication Data
A catalogue record for this book is available from the British Library.

ISBN 1 85567 535 8 (hardback)
      1 85567 536 6 (paperback)

Library of Congress Cataloging-in-Publication Data
Calvert, Peter.
    The South, the North and the environment / Peter and Susan
Calvert.
        p.   cm.
    Includes bibliographical references and index.
    ISBN 1-85567-535-8. — ISBN 1-85567-536-6 (pbk.)
    1. Environmental policy—Developing countries.   I. Calvert, Susan.
II. Title.
GE190.D44C35   1999
363.7'009172'4—dc21   98–50607
                                                                            CIP

CONTENTS

LIST OF ABBREVIATIONS

AIC	advanced industrialized country
CFCs	chlorofluorocarbons
CITES	Convention on International Trade in Endangered Species
CVRD	Companhía Vale do Rio Doce
ECOSOC	UN Economic and Social Council
EIA	environmental impact assessment
ELC(I)	Environmental Liaison Centre (International)
ENSO	El Niño–Southern Oscillation
EST	environmentally sustainable technology
FAO	Food and Agriculture Organization (of the UN)
FOB	free on board
FoE	Friends of the Earth
G7	Group of Seven (leading industrial countries: Canada, France, Germany, Italy, Japan, the UK and the USA)
G8	G7 and Russia
GATT	General Agreement on Tariffs and Trade
GCC	Gulf Cooperation Council
GDP	gross domestic product
GEF	Global Environment Facility
GNP	gross national product
HCB	hexachlorobenzene
HDI	Human Development Index
IAEA	International Atomic Energy Authority

ICIDI	Independent Commission on International Development Issues
ICRW	International Convention on the Regulation of Whaling
IFAW	International Fund for Animal Welfare
IFI	International Financial Institution
IIASA	International Institute for Applied Systems Analysis
ILO	International Labour Organization
IMF	International Monetary Fund
IMO	International Maritime Organization
IMR	infant mortality rate
INCRA	Instituto Nacional de Colonização e Reforma Agrária
IPCC	Intergovernmental Panel on Climate Change
IUCN	International Union for the Conservation of Nature and Natural Resources
IWC	International Whaling Commission
LDC	less developed country
MIC	methyl isocyanate
MVP	minimum viable population
NAFTA	North American Free Trade Agreement
NGO	non-governmental organization
NIC	newly industrializing country
NIMTO	not in my term of office
NNPC	Nigerian National Petroleum Company
OAS	Organization of American States
OCF	*Our Common Future*
OECD	Organization for Economic Cooperation and Development
OPEC	Organization of Petroleum-Exporting Countries
PACD	Plan of Action to Combat Desertification (of the UN)
PAH	polycyclic aromatic hydrocarbon

PGC	Projeto Grande Carajás
PPNN	Programme for Promoting Nuclear Non-proliferation
SAP	Structural Adjustment Package
SIDS	small island developing state
SWMTEP	System-Wide Medium-Term Environmental Plan
TNC	transnational corporation
UNCED	United Nations Conference on Environment and Development (Rio de Janeiro, 1992)
UNCHE	United Nations Conference on the Human Environment (Stockholm, 1972)
UNDP	United Nations Development Programme
UNEP	United Nations Environment Programme
UNFPA	United Nations Population Fund
UNGASS	United Nations General Assembly Special Session
WCED	World Commission on Environment and Development
WCS	World Conservation Strategy
WCU	World Conservation Union
WHO	World Health Organization
WMO	World Meteorological Organization
WRI	World Resource Institute
WTO	World Trade Organization
WWF	Worldwide Fund for Nature (formerly World Wildlife Fund)

CHAPTER 1

INTRODUCTION

How did the environment come to occupy such a central place on the global political agenda? Environmental issues have been important in the advanced industrialized countries (AICs) for a long time. In the UK, for example, conservation groups in the Lake District resisted attempts to dam Thirlmere to provide water for the burgeoning population of Manchester in the late nineteenth century. In the USA, the Sierra Club played a key role in urging Theodore Roosevelt to establish national parks. However, translating these preoccupations to a world stage is another, and much more recent, matter.

Part of the explanation for the development of political interest in the environment lies in the new cultural challenges presented in the 1960s. For Daniel Bell, that decade marked the divide between industrialism and post-industrialism. One of the characteristics of post-industrial society is that, as Bell puts it, the nation-state is too big for the small problems but too small for the large ones. The globalizing tendencies most evident in the expansion of transnational corporations (TNCs) and international telecommunications links began to erode national boundaries precisely at the moment at which many new states were emerging. Thus, problems which transcend national boundaries have become more salient. Further, concern about the impact of unchecked technological development found its expression in the publication of Rachel Carson's *Silent Spring* (1962).

The very idea of the environment was not conceptualized until the 1960s, and there was therefore no environmental politics in the modern sense before that time. Technology itself is also a key feature of developing environmental consciousness. The earth was first photographed from outer space in the 1960s. The striking and poignant view of this 'good earth' taken from Apollo 8, with the dead moon in the foreground, achieved wide coverage at Christmas 1968. It is no coincidence that this was the same time the earth began to be seen as what it is, a 'finite planet with limited capacities to support human life' (Dryzek, 1997, p. 5). This recognition of the limits on the area we all inhabit led to an increased awareness by some of the interconnectedness of all aspects of that world. Environmental science, 'the study of the atmosphere, the land, and the oceans and the great chemical cycles that flow through the physical and biological systems that connect them' (O'Riordan, 1995, p. 2), flourished. However, the complexity of environmental issues, their interconnected and multidimensional nature, is still not fully understood, and still less generally accepted.

Politics is about the distribution of power and the making of decisions in society. Understanding how political decisions are made, who makes them, how they are carried out and whom they affect is essential if the assessments of the environmental scientist are to be translated into policy outcome. The study of the politics of the environment has to proceed while many important issues have not yet been fully resolved, since, if some of the arguments of the scientists are correct, we do not have enough time to wait until they are certain and then start to work out what to do next. We should, it is argued, adopt the 'precautionary principle' and act on fears, not established facts – a practice with which pragmatic and reactive politicians, with their limited constituencies and tenures, are frequently uncomfortable.

Because of the unity of the environment, however, the political problems it presents are especially complex. In particular, they involve two aspects of politics: the politics of each of the 199 or so countries into which the world is divided, which we term 'comparative politics'; and the study of the interactions between their governments, international organizations and non-governmental organizations (NGOs), which we term 'international relations'. This complexity is further enhanced by the role of economic policy, especially that of international capital expressed through the actions of the largest TNCs.

There has been a tendency to see the problems of the environment in terms of management – as a top-down process of telling people what to do. However, politics is not only about structural processes or just about what governments do to people; it is also about human agency as individuals and groups negotiate with wider structures. It is about what people can do to influence government at all levels, business interests and other organizations. The importance of grass-roots participation in protection of the environment is especially pertinent with regard to the 'South'. There has undoubtedly been a strong tendency to see command of the discourse of environmentalism and indeed of the environmental movement itself as products of the middle classes in the 'North', as a result of Northern dominance of international political agendas. But Southern branches of international NGOs, Southern NGOs and local grassroots movements in the South are all now playing very important roles. For example, India now has the largest environmental movement in the world, and this is

Table 1.1 The rise of environmental politics I: pre-Stockholm era

International institutions	NGOs/grassroots politics
IMO	Local and parochial groups (e.g. Lake District conservation groups, the Sierra Club), which provided antecedents for the Friends of the Earth, for example
UN FAO	
ILO	
WMO and predecessors	
UN agencies were by the 1960s pushing environmental issues, but the attention was on industrial pollution in the developed countries	Various national initiatives, NGOs and protest movements in the West in the 1960s, but mainly national in limits and anti-industrial
The only global vision came from IUCN (WCU) and ECOSOC, and called for a more holistic approach and a conference	Some overlap between ecology and peace groups
	The call was taken up by WWF, and interest was mainly in wildlife
The Secretary General was in favour of international action based on multilateral cooperation, bringing together the environment and development (Founex)	Development was largely ignored

Source: based on Caroline Thomas, *The Environment in International Relations*, London: Royal Institute of International Affairs, 1992.

exemplified in local resistance to large-scale dams. Southern groups have better access to knowledge of local ecosystems in the South than their Northern counterparts, and certainly this is the way that the largest Northern groups see it in selecting and working through local groups.

Of course, the unequal size, wealth and status of Northern and Southern NGOs may lead to questions of whether this is a process of co-optation of formerly independent groups by Northern groups already co-opted by the (realities of the) international political system and mass consumption society — what Seabrook (1996) has called 'McWorld'.

DIFFERENT PERCEPTIONS OF PROBLEMS AND SOLUTIONS

Some people simply do not see the environment as important. Their reasons vary. Some genuinely lack interest and knowledge, often as a result of a scientific blind spot, such as that exhibited by Ronald Reagan commenting that 'ninety percent of pollution is caused by trees' (Dryzek, 1997, p. 5; see also Drew, 1981, p. 309; Hertsgaard, 1989, p. 139). For some the environment is a macho challenge, a conquest to be made, a wilderness to tame. Such views are not necessarily badly intentioned: this conquest is often seen in terms of the enhancement of the lot of humankind. They remain prisoners of the 'frontier' mentality, believing in the permanent necessity of human confrontation with a hostile environment. As Dryzek points out, one person's wetland habitat is another's mosquito-infested swamp (Dryzek, 1997, p. 3). Yet others have a boundless belief in the technological 'fix'. They are confident of the continuing capacity of our earth to process wastes or provide substitutes and in the capacity of our species to solve problems. Lastly, for politicians in positions of responsibility or power the problems are so immense, and the contributions to those problems which can be addressed so small because of their incremental nature, that the NIMTO (not in my term of office) perspective takes over.

In short, there are both first-order disputes about the nature of the problems and second-order disputes about the degree to which they matter. Some issues are disputed at the very core (e.g. whether climate change is occurring or desertification spreading, or whether perceived changes are just 'normal' variation). In such disputed areas, there has been some increased funding of research, but this research turns up more and more apparently conflicting evidence, and the certainty which is sought retreats still further. Some changes are agreed to be occurring. There is no doubt that deforestation is happening and there is quite a high level of agreement on how rapidly it is taking place. However, the importance of the processes at work and the morality of trying to stop them are contentious. Some issues raise even greater moral dilemmas and ethical disputes, although the present situation is not disputed. The most obvious example of this is population growth, where current figures are broadly agreed but there are important differences at the level of extrapolations. This is not just because extrapolations differ, but because some see population control as infringing personal freedom of the most basic kind. Some see it as about birth control and therefore evil, and some see it as male power attempting to control female fertility. Of course, for others, it is a purely technical problem about just how many people the planet can feed, house and clothe.

The deep moral basis for such differences (compare other political issues) is expressed in various kinds of environmentalism. 'Environmentalism' embraces a range of views, tending to suggest an anthropocentric view of nature as being primarily something which has value to people and which serves them as a resource. 'Ecology' also varies a great deal, but broadly sees people as one part of an interconnected web of life – a radical form, such as the 'Gaia' hypothesis, suggests that this system is itself responsive to changes and (up to a point) able to correct itself. These moral and philosophical differences are reflected in Southern and Northern perspectives over almost every issue. The South–North debate has been much in evidence at all the international conferences, and markedly so at the Rio Conference of 1992, as will be seen below.

WHAT DO WE MEAN BY THE SOUTH?

'The South' has emerged comparatively recently as a synonym for what was formerly called the 'Third World'. It has gained popularity for two reasons. First, the term 'Third World', coined by the French demographer Sauvy in 1952, has evolved into a concept of development (Worsley, 1967). However, this usage is problematic: levels of development vary so much that the term has different meanings to different people and organizations. Different institutions use different indicators of different types of development. In any case, it is unlikely that any two states will exhibit comparable levels of development on all indicators.

It is true that the term 'South' is misleading in the geographical sense, as can be seen from the map which illustrated the Brandt Report (Brandt, 1980). However, the great merit of the term is that it does not have an attached suggestion of value, or lack of it.

To define the South, as many do, in terms of poverty is very problematic. It is true that even the least developed country of Western Europe, Portugal, is better off in terms of per capita income than any country in Latin America. But this is a single indicator and there are many people in Latin America who have lifestyles which the poor of Portugal – or indeed of an advanced industrialized country such as Germany or the United States – would envy. The global poor, wherever they live, have been identified as a 'Fourth World', but the

more usual use of this term is that of the World Bank, which has since
1978 talked of a 'Fourth World' of the very poorest countries. At the
same time, the World Bank has taken some of the oil-rich Middle East
out of its Southern category. A different problem is posed by the
relatively well developed but still poor newly industrializing countries
(NICs).

However, even with these reservations we can say that the South
shares problems, though the same ones do not necessarily apply in all
the countries of a region, let alone worldwide. Among these are
geography (Bangladesh), lack of infrastructure (Burkina Faso), war/
famine (Somalia) and a burgeoning population in some areas without
economic growth to compensate for the additional demands this makes
on the economy (Africa south of the Sahara generally). Southern
nations may be seen to share economic dependency and a relative lack
of infrastructure. They are socially and economically disadvantaged.
Hence, for the purposes of this book, the term 'the South' will be
preferred and will be taken to mean all the countries of the world not
defined as AICs.

WHAT THE SOUTH HAS IN COMMON

Nearly all Southern states are former colonies. Three exceptions are
China, Thailand and Iran, each of which was subject to considerable
pressure from colonizing powers but ultimately maintained its
independence as a result of conflict between two or more potential
colonizers. A fourth, Ethiopia, escaped colonization in the nineteenth
century only to fall victim in the 1930s to the imperial ambitions of
Benito Mussolini.

However, the historical experiences of colonization in different parts
of the South are in fact very different. They vary with the stage at
which colonization took place and with the economic development of
the colonial power involved, the different policies and practices of the
colonial powers and the nature of indigenous societies. In much of
Latin America there has been a far longer period of independence, and
there was much less traditional society to supersede and/or absorb. In
Asia colonial rule was shorter, independence more recent and colonial
absorption of existing social and political systems much more variable
and sometimes much less complete. 'Protectorates', such as Egypt,
Morocco, Vietnam, parts of Malaysia and Nigeria, were least affected.

Common features of the colonial experience might include: the
establishment of arbitrary territorial boundaries notably in the interior

of Africa; the imposition of a political and administrative order ultimately based on force; and centralized, authoritarian administrative systems. All colonial rule, even that of a democratic country like the United States, which was the colonial power in the Philippines and (briefly) Cuba, is authoritarian. The impact of colonization on environmental politics will therefore clearly vary a great deal, although in general terms the legacy of inequality is the most important aspect, followed possibly by the legacy of non-viable independent states. Both contribute to environmental degradation, as we shall see.

The French Revolution began the break-up of European empires, even before they had been completed. Most of Latin America became independent at the beginning of the nineteenth century, much earlier than the rest of the South. Thus, Argentina was effectively independent in 1810 and formally so after 1816, though it was not recognized as such by Spain until 1853. At various stages in the nineteenth century, between 1806 and 1920, the disintegration of the Turkish Empire created other new states in the Balkans and the Middle East.

Independence came to the rest of the European empires much more recently. The Second World War destroyed the myth of invincibility which helped to make colonial rule acceptable, and this encouraged the growth of nationalism in the South. Often such nationalist movements were led by Western-educated individuals, such as Jawaharlal Nehru in India or Kwame Nkrumah in the Gold Coast (Ghana). After 1945, the will to hold the colonies no longer existed among large sections of the elite of the exhausted Western powers: Britain, France, the Netherlands and Belgium. At this point there was much less contradiction than there had been previously been between the values of the colonial power and the ideal of independence. With independence, however, these perceptions were again to drift apart, as the new state's identity was defined.

However, in all cases the institutions created by the colonial power for its own purposes became the state at independence. This made the newly independent state at once strong and weak. It was strong insofar as it was recognized in international law, intact, functioning and usually centralized. It was weak where divided and ineffective, lacking the structures to hold it together. Only Argentina, Malaga, Mexico, India and Nigeria emerged into independence as true federal states, and in each case the struggle between federal and state governments has gone on to some extent ever since, with varying outcomes. The independent state is also weak in that it is inflexible and subject to nationalist criticism that its forms are inappropriate. It is often associated with the colonial power and subsequently with a small ruling clique, not with society as a whole, and so lacks legitimacy. This lack of legitimacy feeds corruption, which in turn contributes to the lack of legitimacy.

Westernized elites, who see themselves as heirs to colonial overlords, may seek to milk the state for all it is worth. This distorts development. Government does not plan for development, and in any case cannot pay for it. The benefits accruing from control of the state so far exceed those available from other sources that desperation to control the state results in, at best, an undignified scramble which undermines its already tentative legitimacy, and, at worst, the suppression of opposition and the use of clientelism to reward political supporters. The illegitimate state often does not build that legitimacy slowly through evolution. Instead, there is a tendency for frequent changes of constitutions and other superficial attempts to enhance the legitimacy of the state.

Within the weak state, powerful vested interests dominate the political process, making use of it in order to further their own economic ends regardless of cost – a phenomenon generally known as *patrimonialism*. The desire to make use of politics to secure a good job or to protect one's interests in other ways means that political corruption, though by no means exclusively a feature of Southern states, is all-pervasive.

These weak states with powerful vested interests do not serve the environmental interests of the many. Corruption of officials frequently leads to them overriding such limited environmental legislation as weak legislatures succeed in enacting.

Internal insecurity characterizes such states and goes hand-in-hand with external insecurity, which may be summarized as vulnerability due to lack of autonomy. These weaknesses exhibit themselves in the world market and also in the lack of power in institutions like the International Monetary Fund (IMF). The policies of the most powerful Northern countries also impact on Southern states, e.g. US domestic policy in the 1980s, when interest rates at historically high levels increased indebtedness in the South. But Southern states are also much more susceptible to natural disasters, as is evident from the very different capacity to manage flooding in Bangladesh and the Netherlands, for example. Poverty is the root of many of the most persistent environmental problems of the South. Floods, drought and hurricanes all threaten human life and pose serious challenges to governments. However, people are much more vulnerable to these emergencies if they live in inadequate shelter and lack the economic resources to protect themselves against them. Debt will be shown to be a driving force leading to overexploitation of soil and subsoil resources in the process of seeking to maximize foreign exchange earnings.

Redrawing the map of the world to reflect, in terms of relative area, non-geographical variables such as wealth or political power, is, unfortunately, impossible without distorting spatial relationships to a point at which they become completely unrecognizable. However, any map of the world that differentiates countries by, say, their place in the World Bank classification by per capita income is a useful corrective to the simple North–South model, if only because it shows up very clearly the secondary concentration of wealth in the oil-rich countries of the Middle East.

WHAT THE SOUTH DOES NOT HAVE IN COMMON

It has long been traditional to measure economic development in terms of one single indicator, per capita GNP: that is, the gross national product (GNP) of a country divided by its population. By this measure, since 1991 the South itself has been 'pulling apart'. Such measures are full of anomalies. In 1992 there was no real difference between the highest low income country, Indonesia, and the lowest middle income country, Côte d'Ivoire. Owing to a statistical revision, however, in 1994 the highest lower middle income country, Chile, was recorded as having a *higher* GNP per capita than the lowest upper middle income country, South Africa. And not only had Saudi Arabia fallen out of the high income category, but at a GNP per capita of $7510 a substantial gap had already opened up between it and the lowest high income country, Ireland, with a GNP per capita of $12,210, roughly on a par with Israel (World Bank, 1992, 1994).

In 1990 the UN Development Programme (UNDP) published the *Human Development Report*, which used for the first time the Human Development Index (HDI). This ranked countries by a single measure, which included the social indicators of life expectancy and adult literacy and incorporated economic/social aspects in the form of purchasing power. Since then the HDI has been refined to take account of early criticisms, but the basic principle that the general well-being of a country's population can be expressed in terms of a very small number of key indicators has stood the test of time (UNDP, 1994).

The differences within countries can be seen to be as important as, if not more important than, the differences between them. The

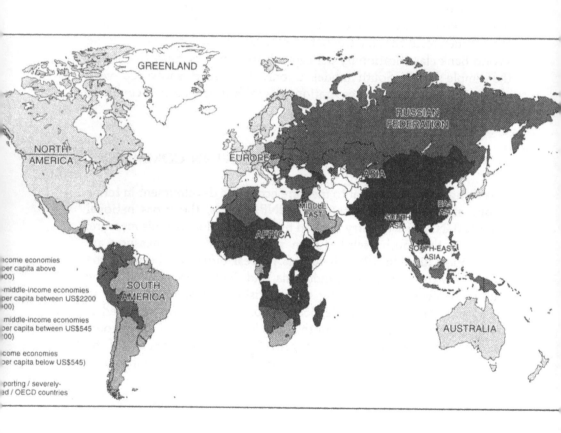

rld by World Bank income categories

percentage in poverty is highest in South Asia and Africa south of the Sahara. South Asia has 30% of the world's population but nearly half the world's poor. Average life expectancy is 76 years in the developed world, 56 years in South Asia, 50 years in Africa south of the Sahara. But these indicators are characterized by wide disparities within Southern states — average life expectancy in Mexico in 1992 was 70, but life expectancy for the poorest 10% was 20 years less than for the richest 10%. The combination of profligate consumption by some and desperate poverty for others has the worst possible consequences for the environment.

The basic fact that determines a country's place in the world is its capacity to feed its people. Unless some other resource is available to be traded for food, the ability to feed the people depends on: (a) the availability of land; (b) its 'carrying capacity'.

Inequality in land ownership is wasteful of huge areas of a vital resource, while putting additional pressure on others. This problem is not confined to the South. Australia, an AIC, has one of the most unequal land distribution ratios in the world as measured by the Gini index, the standard measure of land inequality, but the lack of population pressure reduces the environmental impact. The South is distinguished by the survival of traditional cultivation on the proportion of land available to the mass of the population, and even this is now often under threat. It is important to distinguish between peasants, cultivators who have traditional rights to land, and plantation workers on large estates, who work for wages. Each group faces different threats to the security of what Amartya Sen (1981) terms the 'entitlement' to food and contributes different impacts on the environments.

Peasants have (a) access to land, (b) family labour, (c) small-scale technology and (d) the ability to generate surplus in a cash economy. Even small-scale technology requires cash for purchase and maintenance. Peasants make up some 80% of farm workers in the South. Traditional peasant cultivation was balanced between the need to produce for subsistence and exchange and the need to conserve. The key requirement is production for subsistence. This is conditioned by the nature of the crops available. There are three major variants, based on:

- *wheat* (sub-tropical, temperate), in Europe, Northwest Africa, the Middle East, the 'Southern Cone' of South America;
- *rice* (tropical, humid), in South, Southeast and East Asia, West Africa;
- *maize* (subtropical, dry), in the Americas, East Africa.

Production for subsistence is based on traditional knowledge and understanding of the needs of the soil. It is therefore resilient and, because of its varied nature and limited expectations, forms the best possible protection for the poor against the possibility of famine. The big problem with it is that as population expands the land areas cultivated are subdivided until production is often barely adequate, implying that large sectors of the population must be malnourished. Because peasants are producing large-bulk, low-value crops in the main, they are also not very productive in monetary terms, which leads their contribution to be underestimated by those in the so-called 'modern' sector of the economy.

With rising population, too many people in the South are working too little land. There are two possible 'solutions': *land reform* by redistribution, which is politically difficult given the vested interests involved; and *land colonization*, by clearing and settling hitherto 'unused' land. The latter is of course only possible where land is available and almost invariably has resulted in serious environmental damage (see Chapter 5). Increases in the 1970s in the land area cultivated in Latin America, China, South and South-East Asia have been considerable, though at the cost of damage to marginal land and its fragile ecosystems. The 'carrying capacity' of the land – that is, the number of people a given area can support – has also been increased at least temporarily, as a result of the so-called 'green revolution', involving the use of high-yielding strains and chemical fertilizers. On the other hand, there has been a marked decline in the production of both wheat and maize in Africa, despite (or because of) the increase in production of cash crops for export, such as tobacco and cotton.

The most successful way found so far for increasing the carrying capacity of land is the wet cultivation of rice. Techniques developed thousands of years ago have proved capable of being used over long historical periods without degrading the land: the silt brought down by the irrigation water acts to replenish the soil and the conditions of cultivation helping to maintain neutral acidity. Where conditions are extremely favourable, in Sumatra, for example, the technique produces as much as three crops of rice a year. Two are normal in the south of China, where there is and has been for centuries extensive use of vegetable composts, green manure and both animal and human wastes to maintain and enhance the fertility of the soil. However, the normal pattern is one crop of rice, alternating with other crops such as sweet potato or other vegetables, which may be grown without irrigation during the dry season.

Wet rice cultivation has spread from its original heartland in Asia to

West Africa (Ghana, Nigeria) and Latin America (Brazil, Guyana), not to mention the USA (South Carolina). It requires level, well watered soils which can be rendered watertight by 'puddling'. It is also very labour intensive, although it·can in favourable conditions ensure the necessary security for farmers to grow a wide range of more speculative crops for market. A drawback is that rice production by this method accounts for a significant increase in the production of methane, one of the 'greenhouse gases' (see Chapter 3). Of course, an alternative to feeding the local population is to produce 'cash crops', alternative commodities which can be traded.

Another alternative is to trade manufactured goods for imports of food, the route that many AICs have already taken. Efforts to industrialize Southern states, like other developmental strategies, have had varied success depending on the natural resources available and the techniques employed to exploit them. As in the development of the Soviet Union under Stalin, but with less drastic methods, early attempts to develop Southern countries were based on the traditional 'smokestack' industries of coal and steel. Hence, in Brazil, primacy in the 1930s went to the creation of an indigenous steel industry, centred on the massive Volta Redonda project. In China, during the 'Great Leap Forward', an attempt was made to substitute labour for investment by encouraging the creation of 'backyard' blast furnaces. What metal these succeeded in producing, however, proved to be of such poor quality that the experiment was allowed to lapse, though not, unfortunately, before it had created considerable environmental damage. More recently, competition from the NICs and from Japan has driven down the world price of steel to the point at which there is now a substantial oversupply of basic steel products.

Although there is much less demand for steel than there was, it still remains central to successful industrialization. In the developed world, for a variety of uses ranging from metal window frames to car bodies, it has been superseded by aluminium. Aluminium is one of the commonest elements in the earth's crust, and in the form of bauxite is found throughout the world. However, it is such an active chemical element that its separation from the ore involves a very high input of electrical power. This takes place in a two-stage process, by which the ore is processed into alumina, and the alumina in turn into finished aluminium. Hence, only where there is a very considerable surplus of power available at low cost, as in developed countries such as Canada, which have considerable quantities of bauxite, is aluminium production economic. An additional problem is that the fabrication of aluminium alloys involves relatively expensive techniques if it is to be successful, thus limiting the spread of technology from the AICs into the South.

The result is that only a quarter of world production comes from the South, from Guyana, Jamaica and Surinam. In the case of Jamaica, it is the island's sole mineral resource, and in 1980 Jamaica was still forced to sell most of its unprocessed bauxite for a relatively meagre return (Dickenson *et al.*, 1983, pp. 136–7). The incapacity to exploit resources locally is typical; for example, in Sierra Leone, bauxite is mined by Sieromco, part of Alusuisse of Switzerland. The effect of mineral extraction there has been to create huge areas of devastation.

The high bulk and low value of most minerals exported in this way means that some of the abundant mineral resources of the South have not been exploited at all. Those deposits most readily accessible by river or sea have generally been opened up first, and certainly remain most competitive on the world market: for example, the copper of Chile or Peru, or alluvial tin in Malaysia and Indonesia. However, world economic conditions can change rapidly. The collapse of the world price of copper has dealt a severe blow to the economies of countries as far apart as Peru, Zaire and Zambia. The ability of TNCs to switch production from one part of the world to another makes the negotiating strength of Southern governments much weaker than their nominal sovereignties would suggest. The environmental costs, however, do not follow the TNCs elsewhere. It is left to the South to clean up, or, more likely, not to, since funds to do so are in short supply.

One person in four in the world lives in China. Three out of four live in the South. Over the past 40 years the population of the South has grown exponentially, leading some (including, notably, the Chinese and Indian governments) to place control of population at the centre of their strategy to achieve a reasonable standard of living for their people. The 1992 *State of World Population Report* (UNFPA, 1992) called for 'immediate and determined action to balance population, consumption and development patterns: to put an end to absolute poverty, provide for human needs and yet protect the environment'. According to the *Report*, world population in mid-1992 was 5.48 billion. It would reach 6 billion by 1998, by annual increments of just under 100 million. Nearly all of this growth would be in Africa, Asia and Latin America. Over half would be in Africa and South Asia. This trend is set to continue until at least 2030, and if it does, the effect will be to make Africa even poorer and to depress living standards in parts of South Asia to the point at which the smallest interruption of the economic order would have serious consequences.

Where they have large populations, Southern states also have vast human resources. Many of their citizens have already to be very resourceful simply to survive. But countries, such as those of the East Asian 'powerhouse', that set out systematically to unlock the potential of their citizens by promoting education, particularly for women, are likely to find that their investment is very well spent in terms of the general betterment of society. Nevertheless, the environmental costs of this additional economic activity, if it continues unabated, must be borne by future generations.

SPECIAL CASES 1: OIL-RICH STATES

The diversity of the South at once militates against South–South cooperation and makes for economic complementarity with the North. A special case consists of those less developed countries (LDCs) which have energy surpluses and hence investment capital (Brunei, Saudi Arabia, the United Arab Emirates). Indeed, these and other OPEC members, such as Kuwait, have made development loans available at preferential rates to less fortunate Southern states. Some countries rich in capital but lacking technical skills have made good this deficit by importing LDC nationals, examples being Iraq and Kuwait. The result is that some of the inhabitants of these countries enjoy an exceptionally high standard of living, not just by local standards, but also by that of the AICs. The concentration of energy surpluses and their profligate extraction and consumption present major environmental problems. These countries have yet to discover how to convert their very high energy potential into broad-based, sustainable development, and there are sound technological reasons why it is very unlikely that they will be able to do so (see Chapter 2).

SPECIAL CASES 2: SMALL ISLAND DEVELOPING STATES

The global problems of climate change and the potential rise of mean sea level have, at least for the moment, called attention to the special problems faced by small island developing states (SIDS).

There are two major areas in which such states are to be found: in the Caribbean, in close proximity to larger mainland states, their independence from which they owe to the accidents of colonial rivalry; and in the Pacific, where their isolation gives them a natural geographical identity. At Barbados in 1994, under UN auspices, an organization was formed to defend their common interests. SIDS are particularly vulnerable to natural and environmental problems, owing to their small size, low-lying land and dependence on only one or two export crops. In the long term, their extremely limited supplies of land are easily exhausted. In what is now the Republic of Nauru, mining for phosphate in the colonial period has left one-third of the small island state a waste of dead coral, with its 5000 population crowded into the part that remains. Biological diversity on the remaining land is minimal.

Although as points of access to the maritime environment the islands offer particular advantages, their capacity to absorb significant increases in tourism, their only obvious source of additional revenue, is very limited. As a result of the need to import food and other items for tourist consumption, the island state of St Vincent in the Caribbean actually loses money on tourism. The tourist industry also puts pressure on freshwater supplies. Many of these islands are low-lying and depend on infrequent rains to replenish the water table. These, as in the case of the Maldives, are threatened by fairly minimal changes in sea level.

Dependence on fishing makes their immediate maritime environment particularly sensitive to disturbance. The traditional habit of discharging wastes into the sea, less problematic with a small local population, has to be superseded as a matter of urgency by effective management of wastes and care of irreplaceable coastal and marine resources. There is at once a threat from the tourist influx and a threat to the attractions that bring them. Chief among these are the living corals of which many of the Pacific islands are composed.

Their geological structure, rising from the deep sea bed, means that for energy resources they are for the present extremely dependent on the import of fossil fuels, though in the long run solar, wave and wind power are all practical 'renewable' alternatives. Some of them are so remote that their dependence on the outside world for transport and communication is total.

THE CRISIS OF THE ENVIRONMENT

Given the variety of circumstances of the nations of the South, it is certainly difficult to find common denominators which can define their environmental problems. But the real link is to be found in Gro Harlem Brundtland's statement that 'The environment is where we all live; development is what we all do.' (World Commission on Environment and Development, 1987).

Table 1.2 The rise of environmental politics II: the Stockholm conference

International institutions	NGOs/grassroots politics
Legitimacy given to environmental issues (institutionalization of the issue)	Environmental management was put at the forefront of NGO agendas, establishing a pattern of NGO monitoring that has continued
Linked the environment and development, pointing out the inequity of the prevailing economic order	ELC(I), networking, meetings, discussion of role and influence, a catalyst for change
Led to UNEP (very important), with international contacts in other intergovernmental agencies, and a role in development of SWMTEP and consciousness raising	Provided the basis for critique (by, for example, 'deep ecologists')
The UNEP programme includes: environment and development, environmental awareness, Earthwatch, oceans, water, terrestrial ecosystems, drylands and desertification, health and human settlements, arms, regional and technical cooperation and coordination, fund raising and working with NGOs	
UNEP discussions led to OCF	
The conference pointed up the transnational nature of many problems, bringing together representatives from 114 countries	

Source: based on Caroline Thomas, *The Environment in International Relations*, London: Royal Institute of International Affairs, 1992.

Development at almost all levels, unfortunately, has brought degradation of the environment. The main difference between the AICs and the rest is that the former have already seriously degraded the environment, while the latter seek to do so with as little delay as possible. Degradation of the environment is no new problem: large-scale irrigation causing salination of agricultural land contributed to the decline of Mesopotamian civilization (Glasbergen and Blowers, 1995,

p. 1). But we exploit the environment more intensively now than ever before and the implications of the process today are global.

Goodland (in Goodland *et al.*, 1991) identifies five reasons for believing that the capacity to develop is now reaching its limits:

1. The human economy already uses some 40% of the product of photosynthesis (Vitousek *et al.*, 1986). If, as is expected, the world population doubles before it can be stabilized, it will be pushing the bounds of possibility to feed, clothe and house it.
2. 'The indications of atmospheric carbon dioxide accumulation are pervasive, as geographically extensive as possible, and unimaginably expensive to cure if allowed to worsen' (ibid.).
3. The ozone shield that protects life forms from damage has been conclusively shown to be ruptured and on the most optimistic scenario imaginable will continue to be so until after 2060.
4. Some 35% of the world's land is already degraded. As more marginal land is brought into cultivation, the rate of loss increases, typically exceeding the rate of soil formation by some ten times (Pimentel *et al.*, 1987).
5. There are already so many human beings living on the planet that we constitute a threat to the habitats of countless other species.

Not surprisingly, therefore, the environment is one of the three great issues which Porter and Brown (1996) identify as coming to dominate world politics by the 1990s, together with international security and global economics. International security, global economics and the environment are of course entwined with each other, and likely to become still more so. Depletion and degradation of the environment are already affecting international security, and the impact is likely to increase. The environment is both the source and the consequence of the international system of resource production and use. It is an obvious victim of the liberalization of world trade but also a bargaining chip which will clearly affect South–North relations.

Is the environment a security issue? This goes beyond questions about whether military action over resources which have become scarce is likely, although that will be addressed below. It is really about whether traditional definitions of 'security' are still valid.

The 'sovereignty' of states allows pursuit of any goal without intervention within territorial boundaries. This is even enshrined in Principle 2 of the Rio Declaration, which implicitly admits that states are free to cause environmental damage within their own jurisdictions. The degrees of depletion and pollution of our resources make the environment a global problem requiring a global solution, but 'sovereign' states still seek 'security' for themselves. They do so with a historical legacy that has not favoured the environment. During the Cold War competition, neither ideology stressed the environment. The environment was ignored or exploited. That was the logic of industrialism: the environment was a resource, an input to the industrial process. It is a mark of how far the political agenda has been changed that *The Morning Star* (28 August 1997) has carried environmental headlines on its front page. Not only are the causes of environmental degradation unevenly spread, but so too are the effects. The policy-makers are not necessarily those who experience first-hand the worst effects of environmental degradation at both national and international levels.

A system of competitive, theoretically independent states continues to provide the basis for environmental politics, as it does for the international economic system. And the dominant neoliberal economic ideology externalizes the environment. The 'free market' actually encourages production to seek out areas where constraints are least and costs cheapest. The international economic institutions established under the Bretton Woods system have tended to support these neoliberal ideas in policies such as Structural Adjustment Packages (SAPs), which have encouraged overexploitation of environmental resources (Reed, 1992).

But it is not just those individuals and states with power which have sole responsibility for environmental problems. As Caroline Thomas (1992, p. 7) points out, the South also contributes through misuse of agricultural land, rural overpopulation and urban pollution. Inequality in the South is particularly important. Lack of resources leads to environmental overexploitation by the poorest sectors. For example, land inequality in states such as Brazil leads to the environmental degradation of marginal lands, and to the burning of the rainforest for land clearance.

CONCLUSION

Any solution to the environmental challenges of today has to be sought within the constraints imposed by the laws of the physical and chemical world. However, political conditions will determine what the actual response will be, or, indeed, whether there will be any response. Politicians cannot improve on the laws of science. They can, however, make a bad situation much worse. So our starting point must be with the physical world, with the sources and sinks, and the conversion processes which make life possible.

Table 1.3 The rise of environmental politics III: the Rio conference

International institutions	NGOs/grassroots politics
The end of the Cold War led to the opportunity for further debate, arms reduction agreements and an enhanced UN role	Worldwide spread of green ideas
	Increase in numbers and size of NGOs
	Opportunity and ambiance for other single issue development groups in the South
	Blossoming of protest in the East, environmental political parties in Western Europe
	The importance of environmental disasters and the intensity of exploitation led to a growth in ecology movements (e.g. in India) and responses to perceived threats
	An association with political pluralism
	Development of the capacities of NGOs, with expansion of interests, collaboration with governments and UN agencies, granting of observer status at UNCED

Source: based on Caroline Thomas, *The Environment in International Relations*, London: Royal Institute of International Affairs, 1992.

Second, political action is constrained by the fact that although we live in one world from an environmental point of view, for political purposes it is divided into some 190 nominally sovereign states. These states, moreover, have to interact with one another in a world economic system based on the assumption of a free market in goods. These two assumptions are incompatible with one another, but they give endless scope for procrastination, especially since states are not autonomous and markets are not in fact free. Many observers have noted that TNCs have massive resources, in a considerable number of cases far in excess of those available to the smaller, weaker, Southern states. However, even if TNCs have great power, they have limited objectives. Their purpose is simply to tilt the system in their own favour, and if in the process they cause environmental damage, then that remains the responsibility of the state to check. International action is limited by two factors: international negotiators soon find that they cannot get states to agree to specific actions rather than general principles, and to implement even these vague promises they have to rely on the power and goodwill of national governments.

The political structures characteristic of the South also lead to certain features of societies that condition reactions to the environment and the challenges it poses. In later chapters the following will be recurrent themes:

The economic pressures to use inappropriate technology. Southern states frequently lack the capital to modernize a polluting technology. Despite calls for the transfer of an environmentally sustainable technology (EST) at Rio and since, Northern patents have prevented such transfers. Likewise, the low cost of Southern labour is an incentive to use labour-intensive technology, whether polluting or not. This technology is already in place and all but the largest companies will clearly wish to exploit their existing investment rather than replace it. The local community will likewise prefer immediate returns in terms of jobs and taxes to environmentally friendly deferred benefits.

The perceived costs of environmentally friendly solutions. Dependence on primary industry, with all its consequent environmental costs, is still the starting point, and cleaning up is a cost which Southern governments with their limited tax bases cannot afford. In fact, some of that tax base may be attributable to displaced Northern pollution. Not only do polluting industries deliberately relocate in the South where restrictions are fewer, but they may also use their local knowledge to seek sites for the dumping of toxic waste from elsewhere.

Individual poverty as a major problem for the environment. While individual poverty is a factor encouraging frugality and recycling of waste, it also, encourages degradation of the environment. Poor living conditions lead directly to the organic pollution of water and land. The link between poverty and disease is illustrated by the outbreak of cholera in Peru in the early 1990s, after an absence of more than a century.

The demand for development. Global images of what the North perceives as the good life are flashed around the world and, not surprisingly, the benefits of mass production are sought by people everywhere. We are destined to strive to 'improve' our lot, or, as Brundtland says, 'development is what we *all* do'. Thus Southern states are experiencing an urgent drive to develop, resulting in high rates of deforestation (Jamaica loses 7.2% of its cover every year) and, where land has already been 'cleared', pressures to degrade it by overuse of irrigation, fertilizers and pesticides.

The distorting effects of investment. Investment is consistently represented in the prevailing neoliberal orthodoxy as the way forward for the poor South. Its benefits will, they say, 'trickle down' to the poorest and most vulnerable, who will then not need to exploit their marginal environments so intensively. However, there is little evidence that investment has had such an effect; it has, on the contrary, acted to exacerbate inequalities. It has sought out the so-called 'emerging markets' of East Asia and Latin America, though perhaps with less desirable effects than had been envisaged. No private financial interest has to date identified Central or West Africa as potential 'emerging markets' and begun to invest there. Where international financial institutions (IFIs) are concerned, lending agencies tend to prefer large projects that are least sustainable in environmental terms. A further problem is that where funds have been invested, they may be skimmed off by elites for personal use or by governments for political advantage. Either way, both the poor and the environment are left to bear the cost.

In the rest of this book we shall address, first, the key issue of energy and energy resources in Chapter 2. We shall then investigate in turn each of the earth's three reservoirs: the atmosphere in Chapter 3, the hydrosphere in Chapter 4 and the lithosphere in Chapter 5. In each case we shall examine the sources, the processes and the sinks, the relationship between them and the major political problems presented by each. In Chapter 6 we turn our attention to the biosphere, before considering in Chapter 7 the ecology of human beings themselves, and their relationship to the environment. In Chapter 8 we return to the political questions dividing South and North, and look once more at the factors peculiar to the South that determine its responses to the challenge of environmental change, before concluding with likely future developments and the possibilities of change.

CHAPTER 2

ENERGY

THE SUN

Ultimately, the capacity of human beings to alter or to destroy their environment depends on energy, and all energy except nuclear and geothermal energy comes from the sun. The sun makes a *direct* input to plants through photosynthesis. We make *indirect* use of this energy stored in plants by burning fuelwood. The use of fuelwood continues throughout the South, though there are problems today of finding enough. But modern society, too, rests on the consumption of energy produced by plants. This consumption has been changed and accelerated through technological advances. Plant material can be converted into more convenient forms. But, worldwide, humans have preferred to draw on fossil plants and wood in the form of coal, oil and natural gas.

Nuclear energy exploits the property of certain heavy atoms that they are so large that they spontaneously emit radioactivity, which, in sufficient quantity, will stimulate a self-sustaining reaction and generate heat. However, it still requires a large fossil fuel input to extract the metal, to energize the reaction and to dispose of (or more precisely store) the waste products.

The ideal, in terms of both efficiency and environmental protection, is to exploit the heat of the sun directly. This can be done in two ways. The power of the sun can be harnessed to heat water or another heat-absorbing substance. Houses and cars are warmed by the sun whether we like it or not — the total heat gained in this way in the South is colossal, but most of it is wasted. Not all, however: in Israel and Jordan, rooftop solar collectors already provide up to 65% of hot water, and cheap solar cookers could also economize on the use of fuelwood, especially in Africa, where, despite the abundance of solar energy available, at present fuelwood supplies some three-quarters of cooking needs (Brown *et al.*, 1993, p. 52). Much more efficient use can be made of the sun by encouraging the use of simple passive solar collectors and improving storage techniques. The sun is now being used to heat swimming pools even in the northerly UK, and reservoirs of warm water can be used to heat houses when the sun is not shining.

If light is required, however, we need direct conversion of solar energy to electricity. Discovery of new materials in the 1980s and 1990s has now made this process sufficiently efficient to be economic if the panels are built into new buildings. India, Indonesia and Sri Lanka have plans to encourage the use of solar panels. However, the downside is that production of the complex materials needed for the panels is itself polluting to the (AICs') environment.

BIOLOGICAL SOURCES

Fuelwood

Wood is still the principal fuel for 80% of the world's population, but it is not being sustainably harvested. Quantities have long been considerable: it was estimated in the 1980s that in rural areas of countries as far apart as Tanzania and Thailand consumption of fuelwood was of the order of 1–2 tonnes per capita per year (Brown, 1981). As late as 1971, it was suggested that world consumption of energy from fuelwood was 'greater than that from combined hydro-electric schemes, nuclear power and geothermal sources'. The world then had 'an estimated forest growing stock equivalent to 271 000 million tonnes of coal, to which a potential increment of 7700 million tonnes of coal is added each year'; despite this, only some 13% of the world's annual forest increment was being harvested (Earl, 1975, pp. 6, 43, 53). But the value of fuelwood was not included in the calculation of GNP per capita, giving a false impression both of its importance and of the real level of consumption in developing countries.

There is plenty of evidence from the North about the long-term consequences of excessive fuelwood extraction. Ecological damage from felling of forests became serious in Britain (for example, the Weald) in the seventeenth century and in Sweden shortly afterwards. When Iceland was settled, between 875 and 930 CE, it is estimated that 39% of its land surface was covered by forest; now only 1% remains after cutting for fuel and the overgrazing of the remainder (Morgan, 1986, pp. 171–2). The problem of fuelwood is the problem of forests. Pressure from expanding urban populations has invariably resulted in massive destruction of sustainable forest resources. Today, in Southern countries as diverse as Sudan, Ethiopia, Nigeria, Indonesia and Colombia, as much as 80% of annual timber removals are for fuel. Serious damage has been done in Nepal by deforestation for fuel purposes.

However, wood is the most obvious sustainable source of fuel available and the technology for using it is cheap and easily understood. Immediate gains could be made by introducing more efficient stoves, which burn a fraction of present consumption. The main disadvantage is wood's high water content, typically of the order of 100% of dry weight, which limits its capacity to generate heat (Earl, 1975, p. 23). Secondary forest energy resources include charcoal, methyl alcohol and the gases carbon monoxide, hydrogen and their combinations: producer gas and water gas, generated by spraying burning charcoal with water. Again, care has to be taken with the burning of both wood and charcoal to avoid the risk of carbon monoxide poisoning, and there are other problems. In Ulan Bator, capital of Mongolia, where the temperature falls as low as −35 °C during the long winter and half the 650,000 population live in traditional round felt huts called gers, pollution from stoves burning both coal and wood causes severe respiratory diseases, especially in children. Recently, the UK Foreign Office gave a grant of £35,000 to a local inventor to market an improved stove, which burns a quarter of the fuel and does so much more efficiently (The *Guardian*, 21 April 1998).

Biomass

This term applies to all ways in which energy is obtained by using the product of photosynthesis − hence it correctly includes wood burning.

One alternative already employed in the South is the fermentation of vegetable matter into alcohol (ethanol). This burns cleanly, leaving no harmful waste products, and the main problem is that the crop has to compete with food for land use. In Brazil, a major programme has been implemented to distil sugarcane into ethanol for use in cars and other motor vehicles. A similar programme in Zimbabwe had a common origin in the desire to avoid dependence on imports; the Philippines, too, gets a substantial portion of its energy budget from biomass. Up to 23% anhydrous alcohol can be added to motor spirit without requiring costly engine conversion, thus economizing on scarce or imported fossil fuel, but not gaining the full environmental benefits of burning hydrated alcohol. Brazilian engineers are proud of the fact that they designed and produced special engines to burn hydrated alcohol when they found they were not available from the AICs and by 1992 4.2 million vehicles used them.

Other products can, however, be used. The people of rural India, Pakistan and Turkey burn dried animal dung. About 100 million tonnes of dried dung is burnt annually in India. It would be more energy efficient as well as less environmentally damaging to create digesters to produce gas from mixed waste, but these cost money, and in any case there is a strong argument against using animal dung in this way. Animal dung can be used in more environmentally friendly ways as a natural fertilizer, and its use for combustion is a cost to soil quality.

The use of human waste as a fertilizer, on the other hand, though widespread in countries as far apart as Egypt and China, is a serious health hazard. Collecting sewage for digesting into inflammable 'bio-gas' is therefore doubly beneficial: it improves sanitation and it economizes on fuel use. In addition, burning the methane-rich gas helps to combat the 'greenhouse effect' (see Chapter 3). Digesters are used in more than 43 Southern countries. Their drawback is the capital cost of installation. The technology of the generation of gas and the maintenance of the plant is not complicated, but it needs a boost from the AICs to provide the equipment cheaply.

Human and Animal Labour

The *shaduf* (a bowl suspended by a cord from a long lever) has been used in Egypt since Pharaonic times to raise water into irrigation channels. Simple hand pumps still raise water in rural areas throughout the world, and women and children carry water many kilometres for domestic use. Animals plough fields, help with the harvest, thresh grain and so on, as well as providing a means of transport. None of this is included in the total world energy budget, and the use of human labour in this way is extremely wasteful, seriously limiting the contribution each individual could otherwise make to society.

FOSSIL FUELS: COAL, OIL, NATURAL GAS

Coal

The world's most plentiful fossil fuel, coal, is found very widely throughout the world. However, 57% of the world's known reserves occur in only three areas: China, the former Soviet Union and the USA. Coal has been in general use in the UK since Elizabethan times. The Industrial Revolution began in England because coal and iron were found there in close proximity. Coal use is still steadily rising where high-quality reserves are not found nearby, and this is problematic, because the coal burnt is often of poor quality, burning with high sulphur emissions (see Chapter 3). Coal is also bulky and difficult to transport cheaply except by water. For example, in China most coal is found in the northerly provinces of Shanxi, Hebei, Henan and Shaanxi, while factories and other large users are mainly in the south of the country (Simeons, 1978, p. 31).

Although bulky, coal can be stored in large quantities to even out supply and demand. Most coal has a high sulphur content. Its use, therefore, has significant impact on the environment and was the major cause of primary pollution of cities in the first half of the twentieth century.

The major consumers of coal are also the major producers, and they control the bulk of the world's coal reserves. In 1971, the USSR had the largest known reserves, 56.5% of the world total reserves, and the USA came second with 19.5%. In the same year, the USA produced 30.9% and the USSR 24.9% of world production. China's rate of consumption is a special case, with 24.1% of world production but less than 5% of world reserves. Poland, West Germany and the UK were also significant producers, but with small reserves. Japan imported some 25% of world coal imports, but was already less dependent than on oil (Choucri, 1976, pp. 22–3). But the first oil shock of 1973 made oil generally uncompetitive with coal for the purpose of generating electricity, and plans for conversion to oil were postponed. The actual costs of conversion are not in themselves very high if it is undertaken as part of an overall energy strategy (Hill and Vielvoye, 1974, p. 180).

The recovery of coal presents considerable problems. In the UK, the shallow seams of the Black Country were worked out by the 1920s; in deep mines technical problems increase rapidly. These include sinking deep shafts, increasing the length of driveways, providing face supports, shearers, ploughs and trams or belt conveyors, managing ventilation, salvaging coal faces and managing spoil tips, but all these increase the energy needed to recover the coal. Open-cast coal mining is even more destructive to the environment. Underground gasification, pyrolisis and so on are difficult to control and generate large quantities of carbon monoxide (CO) and other hydrocarbons, which are potentially toxic if allowed to escape. Gas has to be consumed on site. Even in the former USSR, underground gasification could not compete on cost grounds with natural gas or open lignite (brown coal) mining (Simeons, 1978, pp. 169, 185). Liquefaction of coal involves even greater problems. In addition to coal, a source of hydrogen is required. The difficulty of obtaining oil products and the potential dangers to what it saw as its national security meant that the Apartheid regime in South Africa was prepared to bear the high cost of the Sasol programme, a programme to convert coal into liquid fuel. But it also made the plant a very attractive target for a successful sabotage attempt in 1980, and makes little sense in post-transition South Africa (Hanlon, 1986, pp. 17, 74).

Oil

Petroleum is currently the world's most used energy source. Its use as a lubricant began with Drake's oil well in Pennsylvania in 1859, but its modern use as a fuel results from the invention of the internal combustion engine and the discovery of the first 'gushers' in 1901. In the 1940s came a big change in the industry, as production shifted to the Middle East and other Southern states. This was partly because the USA and Russia are the only industrialized countries that have historically been able to draw on large oil resources within their own territory. But another factor was the ability of TNCs to trade off taxes in Southern states against tax concessions at home.

Today world oil production is dominated by TNCs in the form of 'the Seven Sisters', a nickname popularised by Enrico Mattei for the seven (now six) great 'oil majors' which control the production and sale of oil throughout the world (Sampson, 1975; see more particularly Philip, 1982). Of the six, four have their headquarters in the USA: Exxon–Mobil (formerly Standard Oil of New Jersey), Gulf, Socal (Standard Oil of California) and Texaco. One is British, BP–Amoco, and one Anglo-Dutch, Royal Dutch-Shell. 'The industry is highly internationalized and its history illustrates both the importance of competitive pressures in the world market and the ways in which major TNCs attempt to control and eliminate such pressures' (Jenkins, 1987, p. 51).

Two important technical factors help to explain the continued dominance of the TNCs, despite nationalist resentment and the desire of governments to control their own resources. First, prospecting is a skilled business involving advanced technical knowledge, and many more dry wells are drilled than productive ones. The oil companies have the expertise and spread the risk over many different fields. Second, crude oil comes in two main types and varies widely in chemical composition. As no one country is generally likely to produce all its needs from its own resources, an international trade in oil and its derivatives is the inevitable consequence. There are also political factors.

Among Southern states, some have more oil than they can use. Many of the others are entirely dependent on imports. The key questions are: why did production spread to Southern states; why did it become dominated by large TNCs; and why were Southern states initially willing to accept such apparently poor terms of trade? At the end of the 1920s, as abundant supplies of low-cost crude oil became available, threatening the position of the existing companies, consortia were formed to regulate supply in each of the main producing countries: Iraq, Iran, Saudi Arabia and Kuwait. The profits of these joint enterprises were then shared out between the participants. A combination of financial inducements and (in the early days) threats of force induced governments to accept the terms offered. In any case, they had little alternative. Oil was of no value unless it could be extracted, refined and marketed, and they lacked the facilities and the know-how to do any of these things.

Oil has a relatively low sulphur content compared with coal. However, its use gives rise to invisible products of combustion, including nitrogen dioxide (NO_2), which in the presence of sunlight recombine to form secondary pollution in the form of the 'photochemical smog' which was first identified over Los Angeles, California. It is a major contributor to the pollution of the atmosphere in Southern cities and, since leaded petrol is still widely used, pollution by lead is also rising there, while it has fallen substantially over the past 15 years in the AICs.

Gas

Once regarded as an unwanted by-product of oil production, gas was originally simply 'flared off' at the well-head. Gas is, however, found in large quantities on its own. At the end of 1972 proven reserves totalled some 53 trillion cubic metres (m^3), of which one-third were in the former Soviet Union, one-sixth in the USA and another one-sixth in the Middle East. In the USA, gas was developed from the 1930s onwards in its own right, and by 1972 accounted for 33% of total US energy requirements (Tugendhat and Hamilton, 1975, pp. 354–5). However, at the same time 35% of Venezuelan and 60% of Iranian production was still flared off at the well-head. In 1973, only 7.6% of world production moved across international boundaries. The OPEC countries in 1975 controlled only 35% of world gas production, against 65% for oil.

Today, whether piped, bottled or transported in liquid form, gas has displaced oil wherever its use is convenient, but the limiting factor has been transport – only pipelines are really economical. Here the contrast between South and North is particularly clear. Pressure to bring gas into production in Europe when the first North Sea discovery was made at Groningen led to low prices and bulk contracts for power stations, and, ultimately, to the wasteful 'dash for gas' in the 1990s. In the South, pipelines are built instead to facilitate exports. Only when the Carter administration refused in 1978 to pay a fair price for Mexican gas did the Mexican government divert its planned gas pipeline to the Northern frontier to supply its own industrial complex at Monterrey. Bottled gas is widely available in Southern countries, but its high cost limits its attractiveness for ordinary domestic users.

Gas is a relatively 'clean' fuel. However, using it does not solve the problem of the 'greenhouse effect'. As with other fossil fuels, burning gas contributes to climate change by releasing carbon dioxide (CO_2) though burning oil produces some 40% more, unit for unit (O'Riordan, 1995, p. 322). If incompletely burnt because of poor equipment or poor maintenance gas releases the toxic pollutant carbon monoxide also.

The Conversion Problem

Not only are fossil fuels irreplaceable, but burning them to generate electricity is very wasteful. The most efficient thermal plant loses between 60 and 65% of the energy of the fuel in the process of conversion. After transmission and distribution losses are taken into account, the consumer receives only about 30% of the original energy. Hence savings of fuel through restraint in use of power, improved insulation and so on are very great, although in Southern states it is often difficult to persuade governments to set the necessary standards and even harder to get them to fund their use.

Nuclear Power

Nuclear power combines a modern technology, the handling of radioactive substances, with a relatively primitive one, boiling water to make steam. The fuel used is uranium, and the reaction is moderated by the use of lead and graphite. Care has to be taken to avoid contamination of the secondary, steam circuit by the primary circuit which takes heat out of the reactor core.

The USA had the largest known reserves of uranium in 1975, with South Africa and Canada the other main suppliers, and it was traded on the world market at around US $22 a kilo. Electricity generation capacity rose from 8356 megawatts (MW) in 1960 to 39,864 MW in 1972, on the eve of the first oil shock (Choucri, 1976, p. 24). However, the fact that nuclear technology also has important military uses meant that the nuclear powers went to great lengths to ensure that all aspects remained as far as possible under the supervision of the UN-based International Atomic Energy Authority (IAEA).

The nuclear option has been advocated by AICs as a way of controlling CO_2 emissions. Unfortunately, it also gives rise to two very hazardous environmental problems of its own. The use of nuclear reactors at any level involves the generation of both high-level and low-level nuclear waste in quantities too large for present technology to handle. Second, overheating can occur, and has given rise to three major incidents: at Windscale (now Sellafield), UK, 1957; Three Mile Island, USA, 1979; Chernobyl, USSR, 1986. Each of these involved the escape of substantial quantities of radioactive materials to the environment. The Chernobyl disaster, which affected the whole of northern Europe from Belarus to north Wales, effectively put an end to the construction of new nuclear plants throughout the world.

Radioactivity cannot be destroyed and the products of nuclear fission take many thousands of years to decay. Hence storage of the waste products itself is hazardous. The problem is ongoing. Originally, the nuclear powers assumed that they could safely encourage the use of nuclear power by Southern states because they themselves could offer safe facilities for reprocessing and storage. However, this has turned out not to be the case. In May 1997, a chemical explosion at Hanford Nuclear Reservation in California led to the release of plutonium and other toxic substances (The *Guardian*, 28 July 1997). In the UK, an incredible mixture of radioactive materials was tipped into two silos drilled into crumbling cliffs at Dounreay, Britain's nuclear reprocessing plant in northern Scotland; one exploded, showering the plant with radioactive debris. Both will now have to be excavated by robots, and the contents placed in more appropriate storage (The *Guardian*, 1 April 1998). Before this could be done, however, under pressure from the US government, Britain accepted for reprocessing a consignment of fuel rods from the former Soviet Republic of Georgia, which would have the effect of keeping the plant open. The transfer was carried out in conditions of great secrecy owing to the growing hostility of third parties to having radioactive substances crossing their national territory.

The release of radioactivity has been a regular problem with the nuclear industry since it first began, and the Hanford facility, where the plutonium was manufactured for the nuclear bomb that fell on Nagasaki, is already an environmental disaster area. Up to now, governments mindful of their desire to have nuclear weapons have played down the risks associated with nuclear technology, and a substantial industry has grown up offering the vision of limitless, clean sources of energy. The IAEA itself, set up to control nuclear proliferation, was criticized at Rio for encouraging the use of nuclear power and minimizing the associated problems. It has also tended to blur the lines between peaceful and military uses of nuclear energy, since states such as India and Pakistan have been able to take advantage of the availability of nuclear technology on the open market to develop weapons programmes, and in 1998 to conduct nuclear test explosions.

RENEWABLE ENERGY SOURCES

Hydroelectric Power

The principles of water power have been understood since classical times. It was used in Europe from the twelfth century, onwards to perform simple mechanical tasks. In the eighteenth century, water power made possible the Industrial Revolution in Yorkshire and, with the availability of improved materials, water power was harnessed for the generation of electricity as early as 1880s. Today, hydroelectric power accounts for 25% of world electricity supply or about 5% of world's total energy demand, and the vast majority of new plants being constructed are in the South.

Since the 1930s there has been a strong compulsion to build giant dams to maximize power output. Their problems are discussed in Chapter 4 in more detail, but include flooding, resettlement, destruction of natural habitat, sedimentation, interference with fisheries and contamination of aquatic environment by alien species.

Wave and Tidal Power

Wave and/or tidal power could act as an alternative, providing cheap energy for small islands and other remote locations, but neither governments nor the power industry have been prepared to finance the necessary experimental technologies through to development, and engineers still argue about the best way to exploit wave power.

The Rance Barrage in France has been fully operational since the 1970s, generating 240 MW from the tidal surge. There has been strong resistance in the UK to plans to barrage the River Severn in this way. Small-scale tide mills are not efficient, though they could be of value in meeting local needs in the South. Blocking off an estuary in this way has serious environmental consequences, however, and in many Southern states would not only destroy the living of local fishermen and shellfish gatherers but increase the risk of bilharzia and other parasitic diseases.

Geothermal Energy

Geothermal energy is widely available. It sustains the high standard of living of volcanic Iceland. More surprisingly, perhaps, by 1992 Southern countries such as El Salvador got 40%, Nicaragua 28% and Kenya 11% of their electricity from geothermal sources (Brown *et al.*, 1992, p. 37). However, lower energy sources in deep 'hot rocks' are beginning to be exploited for space heating in Southampton, the only place in the UK where suitable rocks underlie a major urban area. Use at present is limited to urban areas because of the high cost of drilling and the need to use heat locally — hot water cannot economically be piped more than 50 km. Strata are usually heavily saline, too, presenting engineering problems in resisting corrosion of pipes and pumps. The technology is not yet readily available in the South, although it might become so in the long term.

THE ECONOMICS OF ENERGY

How energy sources are exploited is very important to their environmental impact. Cost factors include the following.

Production

Initial capital costs are high and production, once started, may incur fixed costs which require production to be maintained in order to avoid, for example, either coal mines or oil wells being flooded.

Transport

In the case of coal, production has to be very cheap to offset the high costs of transport; however, imported Australian and Colombian coal from open-cast mines can today compete on cost grounds with British deep-mined coal in the UK market. The availability of oil depends on refinery and port facilities, as well as a chain of tankers. Oil, being a liquid, cannot easily be stored in any quantity, and oil in transit from well-head to consumer amounts at any one time to only about three months' supply at 1998 rates of consumption.

Depletion

Resources are finite and need to be carefully safeguarded for future generations. However, democratic (and not-so-democratic) governments are under strong pressure to disburse resources to their supporters, corporate or individual. In the case of interruption of supply, voters are inclined to blame their own government. Where the opportunity exists, they will then turn to another source of fuel, which, being unregulated, may well be even more environmentally harmful (burning of fuelwood, peat, dung etc.).

Air, Land and Sea Pollution

Both mining and oil recovery involve significant environmental impact from the initial exploration stage onwards. Tankers collide with other ships and with rocks, and cause serious coastal pollution. Attempts have been made to stamp out the practice of flushing out tanks into the open sea, but the practice continues, frequently where regulation is weakest, in Southern waters. The oil companies can be made to pay compensation, but this does not repair the damage done to the environment.

Power generation results in at least five major air pollutants – fly ash, sulphur oxides, nitrogen oxides, hydrocarbons and carbon monoxide – though it is the leading producer of only one of these, sulphur oxides, largely because where coal is used as the fuel, it is the poorer quality of coal. In addition, the industry produces large quantities of carbon dioxide. It would add only a very small fraction to the cost of electricity to reduce substantially the effects of pollution, but almost invariably utility authorities are most reluctant to do anything of the kind, and in Southern states ageing generation plants are kept going long past their designed lifespan. This is a clear case where a national government itself has to establish a strong regulatory regime to enforce controls on emissions and ensure compliance with regulations. Southern governments are rarely in a position to enforce such a regime, and in any case balk at what they tend to see as inessential costs.

Thermal Pollution

The increased use of fossil fuels is a major cause of global warming, but all power generation produces heat and so contributes to global climate change. In thermal plants, approximately 60% of combustion heat is wasted. Water is the most commonly used 'heat sink', acting as an effective coolant and transporting its heat load to the atmosphere by evaporation. 'Almost one half of all water use in the United States is for cooling and condensing purposes, and the utility industry uses 80 per cent of this volume' (Carter, 1976, p. 28). But the efficiency of cooling depends on the ambient temperature; the higher this is, the more wasteful of water the cooling process.

In the USA, 90% of cooling water is used once and discharged into the river from which it was drawn. The consequences of this have raised sufficient concern for limits to be imposed in many states. Heating rivers may of itself change their chemical composition. But it is much more serious from the biological point of view. Some species like the excess heat and flourish near the outlets. But other species do not. At the very least, the population of the river will be changed, with possible adverse effects for the sustainability of the populations concerned. Cooling ponds can be used, but have to be large if they are to be efficient. Recently, more and more new plants in the USA have been built using cooling towers instead. These work on the car radiator principle, but still use large quantities of water, which may itself be a problem in a Southern country. Much of the water used in wet towers condenses over the surrounding countryside, carrying with it the chemicals which have to be added to it to prevent build-up of scale and organic matter in the condensers (Carter, 1976, p. 41–9).

THE POLITICS OF ENERGY: CONTROL

Energy has been so important to governments that in the twentieth century many have actively sought to control or at least regulate the provision of energy. Others, while professing not to do so, have intervened more quietly. In the USA (with the important exception of Tennessee Valley Authority, set up in the 1930s to provide a benchmark for the cost of electricity), provision has been through privately owned corporations. Between 1917 and 1979, however, many countries sought to meet their energy needs through state ownership and/or control of key resources. Thus, in Britain, the formation of Anglo-Persian (later Anglo-Iranian) as a public/private corporation was deliberately intended to guarantee oil supplies, and the coal industry was nationalized in 1948. In France, the Compagnie Française des Pétroles, also a public/private corporation, was set up in the 1920s and a government monopoly on the sale of oil and petroleum was established (Turner, 1978, pp. 34–5).

Energy being a basic requirement for industrialization, inevitably the search for cheap energy has had to contend with the tendency for large companies to try to gain control of important sectors of the market. TNCs can be very large, and they may operate in many different countries at the same time. The largest have more economic power than many small Southern states. Their scope enables them to play one small state off against another in their own self-interest. Their personnel come from many different countries, and hence are encouraged to replace their individual loyalties with one, their loyalty to the company they serve. The main purpose of the company is to make money for its shareholders in the AICs; it will therefore demand the free movement of capital in order to repatriate its profits, which pass out of the control of the country in which it operates, without necessarily benefiting it in any direct way. Further, they have the ability to arrange their business operations in such a way as to minimize their tax liabilities, and in this process the smaller countries, which lack the resources to police their activities adequately, are the losers.

> Because of inefficiency in the tax machineries of most developing countries and the corruptible nature of most of the poorly-paid tax assessment officials, some multinationals influence tax assessors and end up paying too little tax to the treasury of the host government. In some cases, multinationals evade taxes entirely because of the underdeveloped nature of the tax laws which in some developing countries make the non-payment of taxes a civil, rather than a criminal, offence.
>
> (Onoh, 1983, p. 14)

The large oil companies illustrate the way in which the dynamics of their industry lead the TNCs to alternate competition and monopoly in their attempt to acquire and defend market share (Jenkins, 1987, p. 51). Until the 1960s, the main area of international friction came between the large companies and local monopolies. In the 1970s, they were squeezed out of some markets but their rate of profitability markedly increased. However, since 1979, the balance between regulation and the market has apparently been shifted in favour of the market, and oil companies, both large and small, have been invited back into markets which they had long since abandoned or had to leave. Privatization of utilities, including state oil companies such as Argentina's YPF, and the opening up of domestic markets to competition, leaves the sole purpose of such companies the provision of returns to their shareholders, regardless of environmental or other considerations. The history of the relationship between the transnational corporations and their Southern hosts is illustrative of the power imbalance that characterizes all aspects of environmental politics.

The AICs supported their own oil companies because it benefited them financially, but more particularly because the role of petroleum in the First World War had made them keenly sensitive to the need for guaranteed supplies of oil in wartime. They were right to attach great importance to this fact: in 1941, Japan attacked the USA because of the potential threat to Japanese supplies, and the oilfields of the Dutch East Indies (now Indonesia) enabled Japanese military leaders to continue the conflict as long as they did.

By the 1950s, however, states of the South (Iran, Venezuela) were generating substantial revenue from petroleum and had renegotiated terms in their favour. Russia and Mexico had expropriated their oil industries and brought them under state control. In 1960, OPEC was formed. Between 1960 and 1973, oil production in the OPEC countries increased from 433 to 1307 million tonnes a year, stimulating economic growth, urbanization and an increased sense of grievance. There was grumbling in OPEC about obtaining a larger share of the revenues, but for a time little happened, although in 1970 Muammar Ghadaffi in Libya was successful in forcing a new pricing agreement on Occidental, and his example was followed by others. Then, in 1973, with the outbreak of war between Egypt and Israel, the OPEC countries took advantage of the crisis to force up the price of crude oil, doubling the price twice in the three months before Christmas. Iranian light, which sold for US$2.995 a barrel on 1 October 1973, cost $11.875 on 1 January 1974 (Hill and Vielvoye, 1974, p. 73).

The Shah of Iran warned the AICs that they would have to tighten their belts. 'Some people say that this will create chaos in the industrial world and create a burden for the poor countries,' he said. 'This is true. But the industrial world will have to realize that the era of the terrific progress and even more terrific income and wealth based on cheap oil is finished. They must find alternative sources of energy' (Hill and Vielvoye, 1974, pp. 15–16).

The immediate reaction from the AICs was that the increases posed 'a challenge to the international system that is without close precedent' (Rustow and Mugno, 1976, p. 1). Those who hold this view note that boom conditions in 1973 had in fact left the OECD economies dangerously exposed. They were in consequence plunged into steep recession: average inflation reached 13.5%, unemployment rose to 15 million by 1975. Germany, which acted early, kept inflation in check; Britain, which tried to keep going, did worse than other European countries. This first 'oil shock', when the price of crude oil multiplied fivefold in a year, also sent shock waves through the world markets. The huge increase in the flow of 'petrodollars' to oil producing states created the need for new places to invest the revenue, and semi-industrialized countries such as Brazil and Mexico were high up the list of places that looked attractive. Banks in both Europe and the USA competed fiercely to offer them loans at competitive rates.

The worst affected, therefore, were the smaller non-oil Southern countries ('NOPECS'), which suffered both increased oil prices and rising interest rates. In 1972, oil accounted for more than 10% of import costs only in Brazil; by 1974, 21 countries were in this position. On the one hand, a new dependency relationship was established, centring on the oil countries. OPEC states continued to control the price of oil and refused to give non-OPEC states discounts. With their new wealth and small populations, the oil states also drew in large numbers of workers from the non-OPEC Southern states, whose remittances became a significant element in the economies of their home countries (Hallwood and Sinclair, 1981, p. 71). On the other hand, the non-OPEC states also found their traditional markets severely reduced by the recession in the AICs. The combined effect was a steep increase in the South's debt – a pattern that was to be repeated after 1979.

Another view is more sceptical: 'It is open to debate whether the oil companies deliberately encouraged OPEC to raise prices from around 1968 in order to increase company profits', writes Jenkins (1987, p. 54). 'However, there is little doubt that they have come out of the changes in the international oil industry very well.' Turner rejects this argument, however, on the ground that the Church Commission in the USA, which had similar suspicions, was unable to find any evidence that the companies had even foreseen the possibility of OPEC action, although the Commission did note in 1975 that the companies were still actively helping the producer companies to maintain their oil embargo (Turner, 1978, p. 187). In the AICs there was only a temporary decline in consumption. Led by the USA, the AICs were reluctant to take effective action to curb use, and after 1975 OPEC exports rose again. Libya, where Colonel Ghadaffi had sought to restrict production to keep prices up, was shut out. Meanwhile, the oil companies mounted an all-out effort to find new fields, so that by the end of the decade production was beginning in the North Sea and on the Alaskan North Slope – both areas under the full control of the AICs. And as regards the impact on the non-oil Southern states, it is fair to say that the main increase in debt occurred in a short list of countries (Mexico, Brazil, India, Israel, Korea, Pakistan, Egypt, Argentina, Turkey and Chile; Hallwood and Sinclair, 1981, p. 84), and in this list there were some oil-producing states, suggesting that the increase in debt was at least in part the result of conscious decisions by national governments and/or national ruling elites.

Following the crisis, there was a very marked increase in aid from OPEC countries to other Southern states. However, this aid was highly selective:

> Arab aid has been channelled to the front-line states against Israel on a massive scale, with Egypt, Syria and Jordan having been the main beneficiaries. Some Arab aid has also been disbursed to other Muslim countries qua Muslim countries. Aid to non-Muslim NOPECs has been on a small scale, presumably because the political benefits are viewed as relatively minor by the OPEC donor countries. None of this latter group of countries has received more OPEC aid disbursements than was paid out in increased oil import bills. Despite the often stated view by OPEC members that they are a part of the Third World, it is clear that they in fact have to a large extent differentiated themselves from the rest of the Third World.
>
> (Hallwood and Sinclair, 1981, p. 128).

The increase in oil prices momentarily made the Shah of Iran very rich; unfortunately for him, it meant that his old friends in the USA were slow to come to his aid when his many enemies at last set about the task of getting rid of him. The unrest that in 1979 led to the fall of the Shah briefly interrupted oil exports from that country, and led to their cessation at the end of 1978. The second 'oil shock', however, resulted less from an actual shortage of oil, the overall supply of which fell by only some 5%, than from uncertainty about the future, which resulted in keen competition between the AICs for available supplies. Prices again rose steeply, precipitating recession in the AICs. This reduced demand and hence income for oil states; between 1979 and 1984, oil use declined by some 14%. The impact of the second oil shock, which plunged the advanced industrial countries into recession, sent interest rates soaring as governments tried to control inflation. Since the majority of international loans were made in US dollars, this made matters much worse for the borrowing countries and led directly to 'the debt crisis', which was to dominate relations between South and North in the 1980s.

Meanwhile, at the Tokyo summit of the Group of Seven (G7) in June 1979, leaders of the seven largest AICs had been able to agree a system of quotas and remained reasonably united on the course to follow. A number of proposals to develop alternative energy sources were also agreed, although from the environmental point of view the noteworthy fact is that the immediate response was to develop nuclear power, increase the use of coal and explore the development of synthetic fuels (Putnam & Bayne, 1987, p. 116). However, with the fall in demand, in practice the target figures were never tested. In 1982, oil prices began to ebb, bringing the 'debt crisis' to Mexico and, eventually, to other oil-producing states, such as Venezuela and

Argentina, which had spent their resources rapidly rather than wisely. In 1986, the price of crude collapsed and by 1996, oil was as cheap relatively as it had been before 1973.

This has accelerated the growth of consumption, and by 1999 the world will have reached the point at which more will have been used than remains. An exacerbating factor is that the USA is itself an oil producer: by the end of the 1980s, exploitation of the Alaskan North Slope and the North Sea meant that OPEC no longer controlled a preponderant share of the world's production, and this has resulted in an alliance between the major Middle Eastern oil producers and the US energy lobby to keep up consumption in the USA and, by extension, in the AICs generally. The AICs which are also producers have a strong incentive, paradoxically, to deplete their reserves as rapidly as possible, owing to their relatively high cost of production. But the rising demand in the AICs is not the end of the story:

> Since 1984, the industrialized world which now consumes more than 50 percent of all primary energy, has resumed a path of moderately rising energy demand. A further 24 percent is accounted for by Russia and China. The remaining 26 percent of energy demand comes from the developing world, where the trend throughout the 1980s was a strong rise of 4–5 percent a year, which shows no signs of abatement in the 1990s.
>
> (Tempest, 1993, p. 249).

In 1992 just under 40% of oil came from three countries: the USA, the former Soviet Union and Saudi Arabia. The USA consumed some 25% of the world's oil production (down from about 60% in 1945). Most was consumed as gasoline, diesel or jet fuel – for environmental reasons the previous use as fuel oil is gradually declining. Since refining of crude oil produces many other products, ranging from pharmaceuticals to asphalt, marketing these products is the key to the very high profitability of the oil industry (Philip, 1994). 'Downstream' activities in petrochemicals acted both to diversify their activities and to make them more profitable. In the long run, of course, they have also helped to increase overall demand. The rise in the oil price in the 1970s also spurred the major oil companies into diversification into other fuels, including coal, where by the beginning of the 1980s they controlled 40% of the US coal market.

Few events have interrupted the remorseless rise of oil consumption. The Iraqi invasion of Kuwait temporarily disrupted production in Kuwait, and the UN oil embargo on Iraq has made it difficult (but not impossible) for it to sell oil abroad. The collapse of the Soviet Union has resulted in a decline of consumption across the region. However,

Rio demonstrated that the oil producing countries now formed a formidable lobby in favour of continued wasteful exploitation. The NGOs, in their choice of the delegations that in their opinion had done most to try to wreck the conference, placed top of the list the USA, which had consistently played the most destructive role in emasculating the two Conventions and seven out of ten sections of Agenda 21. In second place was Saudi Arabia, for insisting on striking all reference to renewable energy sources out of Agenda 21, the massive agreed statement not of what was to be done, but of what should be done, which in the event was the main product of the Conference. Third was Japan, the new economic superpower, which consistently gave silent support to the USA, not least in its efforts to safeguard its high level of oil consumption.

It is widely believed in the oil industry that there are still substantial finds of oil to be made in the Middle East, and with the depletion of the oil reserves of the AICs, the North has become increasingly dependent on imports of oil from the South. However, with the collapse of the Soviet Union in 1991, a new factor has emerged, in that the oil industries of Kazakhstan and Azerbaijan, to say nothing of that of Russia itself, have begun to integrate with those of the rest of the world (Tempest, 1993, p. 201). Some predict that falling Russian oil production might make Russia a net importer by 2000; on the other hand, already substantial efforts are being made to modernize the industries of the other states. There is already, too, keen competition for the pipeline routes which will take oil out of the Caspian, avoiding the political instability of the Middle East, and China has made a bold bid directly to Kazakhstan to help to guarantee itself supplies.

The state-centric arena provided by Rio did not succeed in hiding the influence of the TNCs involved in oil production, and NGOs challenged their role in environmental degradation as well. Nigeria provides an illustration of this role and the extent to which environmental costs are incurred without commensurate developmental benefits to those who pay the costs. Oil exploration in Nigeria had begun in 1937 and resumed in 1946 under Shell BP, a joint venture between Royal Dutch-Shell and British Petroleum. In 1956, oil was discovered in commercial quantities near Port Harcourt, from which the first crude was exported in 1958. The prospect of drawing on the oil revenues was an important factor in the decision of Colonel Odumegwu Ojukwu to lead the then Eastern Region into secession in May 1967 under the name of Biafra. The French oil company Safrap (now Elf-Aquitaine) was accused by the federal government of engineering French government support for the secessionist state.

The federal government also exploited the resentment of non-Igbo minorities at Igbo domination. However, once the war was over, the share of oil revenues taken both by the federal government and by other states increased considerably (Onoh, 1983).

After Nigeria had joined OPEC, steps were taken to set up a national petroleum marketing organization. The Nigerian National Petroleum Company was formed in 1977. Within two years, allegations that the oil production records of NNPC and the partner foreign oil companies were incompatible and that huge sums of money had been siphoned off led to a national scandal. The Crude Oil Sales Tribunal rejected the contention that money was missing, but, feeling that NNPC was too big for efficient management, recommended that it be decentralized. This left the way open for the oil majors. By the 1990s, Shell was being criticized openly by the Ogoni for the way in which it had acted in Ogoniland. The Ogoni people strongly resented the way in which their tribal lands were taken over by Shell. They accused the company of having denied their rights and left their land contaminated by both oil and the pollution generated by the new refinery. After very public protests, the activist Ken Saro-Wiwa was arrested by the Nigerian authorities, tried for insurrection, sentenced to death and executed after the Nigerian dictator, General Abacha, had refused all pleas of clemency.

The Brazilian case illustrates the indirect environmental impact of being caught by the 'double whammy' of being a Southern state without oil in the early 1970s. As noted elsewhere, the debt crisis has serious ecological consequences. Brazil's problems stemmed from the fact that, despite many years of exploration, by the end of the 1970s it still had virtually no oil of its own, and was having to pay hugely inflated prices for every imported litre. The military government from 1964 onwards gave top priority to economic growth, which it saw as a matter of national security. Its free-enterprise model for development was dependent on encouraging heavy investment from overseas. When the authorities found that lack of oil had left their country vulnerable to the oil shocks of the 1970s, therefore, they had very little alternative but to grit their teeth and try to bluff their way through. But with the general recession in foreign trade in the early 1980s, it was by no means easy to find anything that would enable the country to export enough even to pay the dollar cost of its oil imports. Brazil's oil imports during the 1980s amounted to US$58.5 billion. Its net foreign debt in the same period increased by $38.6 billion. However there can be no doubt that the oil companies were prepared to sell Brazil all the oil it could pay for, nor that they entered with enthusiasm into the search for oil on Brazilian territory, which was eventually located some 100 kilometres offshore in 1979.

The Brazilian government meanwhile sought a variety of remedies for its predicament. It encouraged the opening up of the Amazon basin to cultivation, encouraged crop diversification and eliminated non-tariff barriers to economic growth, such as trade unions and labour legislation. It stepped up the search for oil, which eventually bore fruit, and embarked on an innovative energy substitution programme, based on the distillation of alcohol (ethanol) from sugar cane. In 1991 US dollars, Brazil invested $10.7 billion in the *proalcool* (ethanol) programme. The Sarney government tried to drop the programme, but was frustrated by its own internal politics; it had successfully created a market for relatively small producers with their own plantations and their own stills, and they successfully lobbied to save it. By the end of 1991, Brazil had saved the equivalent of $20.3 billion on oil imports and was meeting a quarter of its needs for liquid fuel with ethanol (Brazil: AIAA/Sopral, 1992). Although it was by then also meeting a substantial proportion of its petroleum needs from off-shore production, the fact that all the costs of the ethanol programme could be met in national currency gave it a considerable advantage, which, the government argued, was not adequately represented by comparison litre for litre with the spot price of oil FOB (free on board) at Rotterdam (Governo do Brasil, 1992).

The government has argued that the impact of the programme on land use was minimal, since to meet all Brazil's energy needs would have required only some 2% of its cultivated land area. In the long run, sustainable cultivation and distillation of sugar cane could be a useful way of utilizing biomass, although there is a question mark over the question of the disposal of the liquid waste generated, especially in the NE. However, other aspects of the government's attempts to meet its energy deficit, especially the building of giant dams, undoubtedly did have ecologically undesirable consequences, as will be seen in Chapter 4.

Oil-rich states also face the environmental consequences of their production. Before the discovery of oil, the most important economic activity of the Gulf region was pearl-diving. The region was very sparsely populated. Herdsmen moved freely between one part of the region and another; the modern boundaries between states reflect Western conceptions imposed when European states established relationships with the local sheikhs. Most farmers did not own the land they worked: this belonged to the sheikh, and if the farmer failed to deliver his assigned quota of produce he could be summarily deprived of his livelihood. European interest in the Gulf originated from its strategic position on the route to India, and relations with Oman and Bahrain were already established before oil was discovered (Rumaihi, 1986).

Oil was discovered in Bahrain in 1932, in Kuwait and Saudi Arabia in 1938 and in Qatar in 1940, although in each case production did not begin on a substantial level until after the Second World War, and in particular until the Iranian oil crisis of 1951. The discovery of oil offered the rulers of these states a tantalizing prospect. With so much wealth at their disposal they could not only buy off political opposition but also offer their citizens a standard of living equal to that of the most advanced industrial countries. There was therefore no incentive to restrain production and, not surprisingly, most of the revenue generated from oil went, and still goes, into current expenditure, including the provision of housing, schools, medical services and the like. Until 1973, only limited economic development took place in any of these states and none elsewhere in the region (El Azhary, 1984, p. 3). Instead, they developed a 'leisure society' in which male citizens felt little or no incentive to work.

Development after 1973 took two main forms, both constrained by the shortage of labour. The first was to expand downstream oil activities. Refining as such could not profitably be expanded, as existing capacity was already enough to cope with all foreseeable needs but plant capacity for the liquefication of natural gas could be expanded in Saudi Arabia, Kuwait and the United Arab Emirates (UAE). Until 1966, when a fertilizer plant was established in Kuwait, there was no local petrochemical industry. Here too, expansion was constrained not by environmental considerations but by lack of demand. The main possibilities for expansion in the non-oil sector, in iron and steel, aluminium, copper and cement, all depend on the continued availability of cheap supplies of energy to offset the high costs in other respects. In the case of aluminium, this includes the need to import not only bauxite from Guinea but almost all other necessary materials except petroleum coke (Kubursi, 1985, pp. 44–61). Hence, the diversification of industry would actually have the paradoxical effect of accelerating the depletion of indigenous reserves of oil and gas, and doing so at a very high environmental cost.

It is still arguable whether or not the option of economic development in the Western sense was actually open to the Gulf states. Certainly it did not take place. The central problem was that, with the exception of energy, almost everything had to be imported. In the past, scarcity of freshwater had meant that the region produced only just enough food for the subsistence of its very small population.

With the expansion of oil production, the agricultural sector ceased to be able to attract or retain labour and even with heavy subsidy was unable to compete with imported products. The few farmers that are left work part-time, and in many cases for their own interest rather than for an economic return. Most meat, cereals and fruit are imported. Fish stocks in the Gulf were already depleted by overfishing before both fish and bird life was seriously threatened by the pollution of the shallow, enclosed waters during the Gulf War. No significant indigenous agriculture-based or food-processing industry could therefore emerge.

Last, but not least, the Gulf states came to rely on the services of immigrant workers to an extent unparalleled anywhere else in the world, partly at least because half the available labour force – namely women – was not considered eligible for employment. Initially, many came from other Arab countries, but after 1973 an increasing number came from Sri Lanka or South-East Asia. In Kuwait, the number of foreign-born workers exceeds the native population, and in Qatar and the UAE, the local population is a third or less of the total. 'Rules and laws are applied to the immigrants that are, at the very least, inhumane' (Rumaihi, 1986, pp. 46, 126).

In short, the Gulf states have become a highly artificial society entirely dependent on the over rapid depletion of an irreplaceable natural resource to compensate for the shortcomings of their environment. Such development as has taken place is unsustainable. Clearly, energy alone is not a satisfactory basis for sustainable development.

THE INTERNATIONAL POLITICS OF OIL

The size of large companies often invites comparison with that of governments. However, the question of how much influence they exercise (a) at home and (b) on host governments is very difficult to answer. The literature is polarized and unreliable.

Turner notes that from the beginning the oil companies were on the defensive. Exploration was problematic and start-up costs were high, but governments were not slow to recognize the value of the product and to seek to gain from it. The Standard Oil home monopoly was broken by the US courts in 1911, Russia expropriated the oil companies in 1917 and in the post-war period state industries emerged in Latin America. By the 1940s the Venezuelan 50:50 company/government share principle was generally established. Between 1947 and 1970, the price of Middle East oil fell, while the share paid to Saudi Arabia went up fivefold.

The oil majors have been accused of promoting Western intervention in coups, revolutions and wars. However, there are other reasons why intervention took place, namely the Israeli issue and the Cold War. In most cases there is evidence that companies had to accept overriding political/strategic imperatives, but that countries and companies colluded with one another because they had a shared interest in maximizing production, as the following cases show.

- *The fall of Mussadiq in Iran.* Mussadiq's overthrow was engineered by the US Central Intelligence Agency because it feared Iran would 'go communist'. Mussadiq had nationalized Anglo-Persian (Anglo-Iranian) holdings, and a boycott of Iran by the other oil majors was significant in weakening his position. Afterwards, Anglo-Iranian (now BP) gave up 60% to Shell and the US majors and received substantial compensation from them (Philip, 1994, pp. 118–19).
- *The Cuban Revolution.* The majors were told by the US government not to refine oil supplied to Cuba by the Soviet Union. Ambassador Bonsal confirms that the companies themselves were resigned to refining under protest and seeking redress in the courts. In fact, the companies had actually made advance payments to the new Cuban government without complaint. However, the ban was enforced and resulted in the nationalization of all US companies in Cuba. Ironically, a few years later, for the convenience of the oil companies, Cuba was being quietly supplied from Venezuela, while Soviet oil was being delivered to Spain.
- *Civil war in Angola.* Gulf continued to work with the MPLA government in Angola until conditions in the oilfields became too dangerous. In February 1976, at the prompting of the State Department, they resumed work with the new regime. Given the recent role of OPEC at this period, it was natural that both the US government and the producers would wish to safeguard other sources of supply anyway.
- *Bolivia.* Gulf was close to the Barrientos government (1964–9), and the record shows that payments of doubtful propriety were made to it. In 1969, the Ovando government expropriated Gulf's properties; subsequently the company accepted the compensation offered. Ovando claimed later that Gulf aided his overthrow in 1970, but it would hardly have supported the extreme left Torres government that followed. In fact, events during this period can be fully accounted for by the known fact that the military governments of

both Brazil and Argentina were active in Bolivia, promoting their own national interests (Turner, 1978, pp. 68–79).

Similarly, Western governments' interest in oil has often been overstated. Turner argues that the only case since 1945 when security of oil supplies was a major consideration in UK policy was the Suez fiasco of 1956. In the USA, the evidence is that domestic oil was of great political importance but international oil was not. 'The picture is one of an industry whose influence on foreign policy has been rather less than its size would indicate' (Turner, 1978, pp. 102, 106) In fact, before 1973 the State Department had run down its oil expertise, and US oil production had fallen to the point at which it was not able to make up the deficit caused by the OPEC challenge (ibid., p. 56).

A key problem in North–South relations, however, remains the difference in perceptions between the two on who should have entitlement. As Reid notes:

> The problem is well illustrated by the words of Gerald Ford, president of the USA (1974–76), then as now the nation with the highest per capita consumption of oil: 'No one can foresee the extent of the damage nor the end of the disastrous consequences if nations refuse to share nature's gifts for the benefit of all mankind' ... Sadly, he was referring, not to America's reluctance to be content with its fair share but to the 'selfishness' of the oil-exporting countries.
>
> (Reid, 1995, p. 16).

There are also examples of South–South conflicts over energy resources. The example of the Chaco War (1932–7) between Bolivia and Paraguay is often cited. An important factor in the struggle for the Chaco plain (in present north-west Paraguay) was the belief that under it would lie an extension of the southern Bolivian oilfield. However, Paraguayan forces stopped at the far edge of the plain, within sight of the oilfields, and did not occupy them, as they could have done, and so far no oil has been found in the area they did gain. Moreover, in practice strategic considerations were much more important: the fear that Bolivia would pursue its extreme claim by occupying the Chaco Boreal and establishing garrisons on the west bank of the Paraguay River made resistance to Bolivian pretensions a matter of national survival, and hence was a significant factor in the lead up to the war (Wood, 1966).

More recently, the Venezuelan navy was deployed in the Gulf of Maracaibo to check Colombian claims to the off-shore fields, and the coastal boundary has not yet been finally delimited. The possibility of substantial oil finds also fuels two of the most intractable disputes between Asian states, the coastal boundary disputes between Thailand, Cambodia and Vietnam in the Gulf of Thailand, and the contest between China, Taiwan, Vietnam and the Philippines for the Spratly Islands and the coastal zones around them. Neither can in the end be resolved without an international conference, but in the meantime China has on more than one occasion used force to reassert its claims after other states have tried to make theirs (Morales, 1978, pp. 76–94). A number of production accidents in the region have spurred the adjoining states to take measures to combat marine pollution by oil, which threatens fish stocks, their most important protein source (*ibid.,* pp. 111–13). Last, but not least, it was the sense that Kuwait was undercutting the price of oil and his urgent need to recoup his country's finances after the eight-year war with Iraq that led Saddam Hussein to order the Iraqi invasion of Kuwait in August 1990 (Freedman and Karsh, 1994).

Today, the two most striking things about oil as a commodity are the extent to which it is traded and the fact that oil is produced in LDCs (and former LDCs) and consumed in the AICs. 'The main reason why the oil trade is so heavily structured around oil exports from these LDCs and ex-LDCs to the major industrialised countries is that most First World oil producers consume most of (or more than) their national output. Norway is the only really significant First World oil exporter' (Philip, 1994, p. 3). The reverse is the case in the former Soviet Union, since Russia is a major exporter to the other former Soviet republics. China consumes all its own output, and in 1997 concluded a very significant deal to take a substantial part of the production from independent Kazakhstan in return for inward investment of £6.3 billion in its Caspian oilfields (Meek, 1997).

The universal use of oil does, moreover, make the oil companies a very powerful lobby against any restraint on the consumption of oil for environmental reasons.

CONSEQUENCES OF UNREGULATED EXPLOITATION

Although reserves (oil and/or gas that is known to exist at depths capable of being exploited by known technology) have increased and energy efficiency has improved, more energy is being used, and resources (all the oil that there is in the world) are finite. Massive population increase will inevitably lead to demand for greater use of energy – the world population will more than double by 2150. Reserves of oil were estimated in 1950 as 18 years (i.e. until 1968) at the then rates of use. By 1989, however, new reserves had been located and were expected to last 44 years (i.e. until 2033). The comparable figure for gas was then 58 years (2047), and for coal 220 years (2209). But this is again at current rates and disregarding existing population trends (see Chapter 7). Despite increases in energy efficiency after 1973, 125 countries had increased their total energy consumption by 1987, and world energy use had gone up by 20% (Harrison, 1992, p. 277).

It would be extremely foolish to deduce from this that if more oil is needed it will inevitably be found. 'It should be remembered that substantial conservation efforts require a long time. For example, reducing by half the amount of energy needed for producing ammonia has taken 60 years. The replacement of one form of energy by another as the dominant one is a long historical process too. It took 70 years for oil to move from producing 4 percent to providing 44 percent of the world's energy'. (Noreng, 1978, p. 23). Southern states, therefore, apart from the select club of oil producers, should assume that they will have to continue for some time to live in dependency on the North and the oil majors.

By 1999, the Maguire Oil and Gas Institute argues, we shall have passed the mid-point of depletion of all known reserves (www.cox. smu.edu/maguire/Environment.htm). The really difficult question is whether new reserves will really become available, where they will be found and how much more expensive – in all senses, not just financial – it will be to get at them. In the case of oil, there must, in any case, be a theoretical limit to how deep drilling can go before the rise in temperature makes further discoveries chemically impossible.

The modern world economic order is to a quite remarkable degree dependent on transport. Until the Industrial Revolution, most things were grown or made close to the point at which they were used or consumed. Water transport was slow but energy-efficient. The classic case of 'rational' planning leading to a country's dependence on internal transport over long distances was the former Soviet Union; its disintegration has shown how expensive this mistake was. Southern states are dangerously dependent on access to markets in the AICs; as falling world prices show, only in energy are the AICs so dependent on the South, and then on only some very small parts of it.

Arguments that there will always be more at a price take no account of the social consequences. In those Southern states that have become accustomed to living well on the proceeds of depleting resources, there would be unrest long before resources actually ran out, unless change is managed. And then what? Solar energy can be used for land transport, and wind and water have been used successfully. But so far only hydrogen has been talked about as a possible substitute for kerosene as a fuel for aircraft. Liquid hydrogen is bulky, needs strong containers, is very inflammable and consumes a great deal of energy to produce. And petroleum is also an essential raw material for synthetic chemistry to substitute for minerals, so using it as fuel is something that our descendants will find incredible, if they can afford the time to think about it.

Even in 1979, at the height of OPEC's apparent power, one author noted:

> Certain small countries – which are now rich in petroleum – have already started thinking about the adaptation of their economy to the time when petroleum will disappear not only as a natural resource but also as the only source of the national revenue. Petroleum and natural gas may preserve their supremacy over the other power resources up to the end of this century. However they are not only power resources but also sources of very important raw materials which are used in various other ways – hence their tremendous exploitation at a constantly increasing rate.
>
> (Elian, 1979, p. 19).

Elian also notes that, nearly a century before, Mendeleev had said that burning oil was like burning banknotes (Elian, 1979, p. 20) – meaning that oil was such a valuable raw material that to waste it in this way was absurd. The tragedy of the present situation is that almost every other environmental problem relates in some way to the present use of oil, and will require the use of energy in some form to correct it. As long as energy is wasted in a way that actively harms the environment, all the other problems will continue to get worse, and the South, with its smaller resources, is least well placed to do anything about it.

CHAPTER 3

THE ATMOSPHERE

The atmosphere is the smallest of the earth's geological reservoirs in terms of mass, but the one on which we are all, North and South, most urgently dependent in order to survive. Its limited mass makes it particularly vulnerable to contamination, from both natural and artificial causes. Gases and vapours mix quickly, which favours the rapid spread of contamination but at the same time ensures that it is speedily diluted. Overall, some four-fifths of the atmosphere consists of nitrogen (78.084%), one-fifth oxygen (20.946%) and one-hundredth argon (0.934%), with varying traces of water vapour (typically 0.4–0.5%) and a very small amount of carbon dioxide (0.00036%). However, although the composition of the atmosphere is fairly uniform, there are parts of it that are not as well mixed as others.

Since warm air rises, the lowest part of the atmosphere, the troposphere, on which South and North alike rely for air to breathe, is well mixed. Convection currents carry air warmed by contact with land or sea upwards over the tropics, drawing in cooler air from the polar regions.

At a height of some 15–25 kilometres, in the tropopause, this effect ceases. Air above this zone is warmed directly by the sun, so that the coldest air lies at the bottom and a system of layers builds up, which gives this region the name of stratosphere. In the turbopause, above some 120 kilometres, there are so few molecules that gravitation causes the heavier ones to settle to the bottom. Here, very small amounts of contamination can have a serious effect, and because there is very little mixing it takes a long time for it to be dispersed naturally (Andrews, *et al.*, 1995, pp. 12–18).

The *residence time* of a gas – the length of time a typical molecule stays in the air – depends on how reactive it is and how soon it comes into contact with something with which to react. Because of the mixing effect in the troposphere, most residence times are typically measured in days. However, in some cases they are very much longer. The residence times of the two most important 'greenhouse gases' (see 'Climate change', p. 61), carbon dioxide and methane, are 4 years and 3.6 years respectively. And nitrous oxide (N_2O), an industrial pollutant, will remain in the atmosphere for between 20 and 30 years

(Andrews, *et al.*, 1995, p. 29). There are also quite a lot of fine solid particles (particulates) in the air, but all but the very smallest of these are quickly washed out by the precipitation of water vapour as rain, hail or snow.

The first point that has to be made is that there is no 'quick fix' to the problem of atmospheric change or pollution. Politicians have to recognize that the laws of physics and chemistry alike lock the problems of the environment into a much longer timescale than they are used to considering, and the 'lead time' for decisions has to be that much longer. Indeed, as we shall see, climate change is now already certain to take place; the only questions left to be answered are: can it be managed so that the least harm occurs; and can we stop it from getting worse in time to ensure that our successors can cope?

CARBON DIOXIDE AND METHANE

The main greenhouse gases, the increased emissions of which cause this rise in mean world temperature, are carbon dioxide and methane.

Carbon dioxide (CO_2) is a natural constituent of the atmosphere and is, indeed, essential for plants, which reprocess it into plant matter and oxygen. The most important human impact on the atmosphere is the emission of carbon dioxide as a consequence of the burning of fossil fuels and the burning off of a huge quantity of biomass in the form of the world's forest cover, especially in the tropical South. Some 8.2 billion tons of carbon dioxide is being given off into the atmosphere each year. This is thought by environmentalists to contribute more than half of the global warming effect, and more than two-thirds of the global production of greenhouse gases is due to the burning of fossil fuels. The concentration of carbon dioxide in the atmosphere is increasing greatly – it was 325 parts per million (p.p.m.) in 1970, and 361 p.p.m. in 1998 – despite the fact that each year some 1.6 p.p.m. is either absorbed by the sea or photosynthesized by plants. Our attention is drawn to the significance of this by Woodwell:

> We have over the course of the past century increased the carbon dioxide content of the atmosphere by approximately 30%. Over the past 260,000 years, we have a record from ice cores, showing a correlation between carbon dioxide concentration in the atmosphere and temperature. Over that period of 260,000 years the carbon concentration did not exceed 284 parts per million (p.p.m.) at any time. The present concentration, as a result of human activities, is about 360 p.p.m. and rising at a rate of about 1.5

p.p.m. annually. That's a very significant change. Projections of future changes are for at least that rate or higher rates of accumulation.

(Dr George Woodwell, in Epstein *et al.*, 1996).

Methane (CH_4) is a natural product of the decomposition of organic matter. Although it was (and is) very small in quantity in the atmosphere (in 1990, approximately 1.7 p.p.m.), atmospheric methane is a very effective greenhouse gas, and levels (which are estimated to have been only 0.8 p.p.m. 150 years ago) are rising by some 1% per annum. Rice (paddy) fields are a major source. So are cattle. Hence, the conversion of tropical rainforest to poor quality ranching land, as in Costa Rica, has a double effect. Cattle, being ruminants, are able to break down cellulose and so by enteric fermentation generate far more methane than goats, pigs or human beings. A great deal of methane is also held in suspension in the permafrost and the sea bed, and much of this is likely to be liberated if global warming takes place, thus accelerating the effect. Fortunately, atmospheric mechanisms exist for the complete oxidation of methane to carbon dioxide, but this will take time and may, in any case, not solve the problem.

CLIMATE

Natural influences on climate include the incidence of solar energy, circulation of water and air, continental drift, patterns of plant life, volcanic activity and forest fires. Human influences include the size and distribution of human settlement and land clearance.

The main engine of climate is the warming of the tropical regions by solar energy, causing the warmed air to rise. Most Southern countries lie in the tropical and/or sub-tropical zones. However, both Chile and Argentina, which are southern in the geographical sense, span the entire range of climates from sub-tropical to sub-Antarctic. Outer Mongolia and much of China fall within the northern temperate zone.

In tropical countries there is little division between hot and cool seasons, but there are other very important climatic differences between countries and between regions. For example, West and Central Africa has a hot wet equatorial climate, very heavy rainfall and a rapid rate of evaporation. East Africa has a wet tropical or dry savannah climate owing to its mountain and plateau features. The Horn of Africa is arid, though many of its problems are exacerbated by human action (or inaction).

A vertical separation of climate zones is often more important for most purposes than a horizontal one. Altitude is the key to the nature of human settlement in such regions. It is essential to the growing of cash crops such as tea and coffee. Not only do these crops need cooler conditions, but the height – preferably with forest cover – is required for reliable precipitation. The felling of trees can have major climatic effects at any altitude. For example, in Panama it has been accompanied by prolonged dry spells, which have in turn made necessary restrictions on the draught of ships and the use of locks, and hence on the number of ships that can pass through the Panama Canal (http://www.pancanal.com/advisory). Low-lying islands such as those of the Caribbean or the Maldives lack both forest cover and moisture, and many of the smaller ones are uninhabited in consequence. The Bahamas rely on imported water.

Within the tropical zone, the shifts in prevailing winds bring the monsoons, seasonal winds that carry large quantities of moisture with them; hence local reliance as in the western Ghats of India or in Sri Lanka, on the storage of water from the monsoon in 'tanks' to use during the dry season. The complexity of the patterns involved, even in a small area like Sri Lanka (which is about the same size as Ireland), is shown by the fact that the north-east (mid-October to mid-February) and south-west (April to June) monsoons affect different parts of the island; part of the island, notably the main city Colombo, gets both, and other parts get neither and are arid. Geologically, the monsoons are a relatively recent phenomenon, owing their origin to the orogenic processes that created the Himalayas and so blocked off the access of warmed tropical air to central Asia. The emergence of the Andes has had a similar effect on the climate of South America, causing Patagonia to cool drastically.

The seas, too, are not uniform. Their influence on the climate is evidenced in the most striking way by the phenomenon of El Niño, nowadays known to scientists as the El Niño–Southern Oscillation (ENSO). The term originated on the Western coast of South America, where normally the icy Humboldt Current flows northwards from the Antarctic close to the shores of Chile and Peru. From time to time, however, a current of warm water originating some 4900 kilometres out in the Pacific shifts its position; since this usually occurs around Christmas time, the Spanish-speaking locals named the phenomenon after the Christ child. The consequence that concerned them was that fishing was disrupted, as the Humboldt Current was able to move westwards and no longer brought good catches. But in recent years it has been shown conclusively that this change – the causes of which are still unknown – not only causes hurricanes in the Caribbean and drought in the western USA, but affects the climate of the whole world, delaying, for example, the onset of the monsoon in South-East

Asia. Southern countries, moreover, are much more seriously affected by such changes than the AICs, as they have much less capacity to respond to the challenge of natural disasters.

El Niño illustrates the interconnectness of environmental change. Natural disasters may be the consequence of the changes in ocean currents affecting climate, but it is widely surmised, though of course unproven, that the increasing frequency of El Niño itself may be a result of climate change. The impact in 1998 of Hurricane Mitch is said to have set back the development of both Honduras and Nicaragua by 30 years.

CLIMATE CHANGE

Some 4.5 times as much energy was consumed in 1985 as in 1950 as a result of increased economic activity. The rise in per capita energy use has slowed, reflecting the situation in the developed world, but it is now accelerating in the South and is likely to do so further as industrial and domestic use both expand. The industrialized world's share of fossil fuel consumption, for example, has declined to less than 50% of the world total, while consumption in developing countries has risen by a factor of four in 30 years, and in the transitional economies of Eastern Europe by a factor of 2.4. The consequence has been a massive increase in emissions that contribute to what has become known as the *greenhouse effect*.

The greenhouse effect is the phenomenon whereby the earth's atmosphere traps the radiant heat of the Sun. This makes life on earth as we know it possible, since if heat were not retained in this way the mean temperature of the earth would be below freezing point. But in recent years concern has been widely expressed that the quantity of greenhouse gases being produced by the burning of fossil fuels, slash and burn land clearance, paddy (rice) cultivation, cattle raising and so on is raising global temperatures, with unknown consequences for the stability of the world's climate.

The problem of global warming (climate change) was first given official voice by the Intergovernmental Panel on Climate Change, a group of 700 experts given the job of monitoring and projecting change by the United Nations in 1987. Their best guess is that *rates* of temperature and sea level rises will increase by three to five times in the next 100 years.

Until 1997, 1995 had been the warmest year on record. 1997 set a new record with an average surface temperature for the planet of 0.43°C above the 1961–90 average of 14°C. On present projections, the WMO expects 1998 to set a new record, with an average of 14.58°C and an increase of 0.58°C compared with the 1961–90 average (The *Guardian*, 18 December 1998). Recent research published in *Nature*, drawing on tree ring, polar icecap and other evidence, has shown that average northern temperatures in 1990, 1995 and 1997 were the highest experienced since 1400. 'A large positive correlation' was found with rising levels of carbon dioxide and methane in the atmosphere (The *Guardian*, 23 April 1998). The ten hottest years in recorded history have all occurred since 1983. There are, of course, conflicting opinions about whether this situation arises from human activity or natural variation, but there is no doubt that the level of carbon dioxide has gone up. Generally, global warming is expected to bring higher sea levels in the longer term, as well as severe climate disruption, accompanied by floods, droughts and hurricanes. Such catastrophes have a disproportionate effect in the South, where the infrastructure is least developed and where people are most vulnerable. So at the least there is a strong case for not increasing carbon dioxide levels unnecessarily.

An awkward problem is that climate change will not affect all countries equally. Global warming will actually benefit some in the short term, since rainfall increases will not be uniform. However, changes in the flora and fauna of the planet are inevitable. Some non-tropical forests will be lost, some species will become extinct and agricultural production will shift northwards. Climatic zonal shift will be devastating for many animal and plant species, owing to the lag between climatic change and species evolution. Migration, so easy for human beings, is much slower for small animals and very slow indeed for most plants. Even birds have problems — the capercaillie, for example, threatened by the warming of its Scottish habitat, has no obvious place to go. Even the most conservative estimates of the likely effects of global warming imply a rate of change that would be at or beyond the limits of what species have hitherto found it possible to accept. Southern states appear much less able to cope with such drastic changes than Northern ones.

There are particular problems, which could potentially have a negative effect on the South, in northerly latitudes, in Russia, Canada and Alaska. There, in the past, it has been possible to assume that the soil would remain permanently frozen.

Frozen soil that underpins much of Alaska is melting, causing roads to collapse and landslides, and revealing that the United States may be suffering one of the first tangible consequences of global warming. Scientists have discovered that temperatures in Alaska are rising three times faster than in the rest of the Arctic. Three feet below the surface, the permafrost soil that is supposed never to thaw is one degree warmer than it was a decade ago. Prof John Walsh, an atmospheric scientist at the University of Illinois, said: 'If the permafrost melts, it becomes a soggy surface. It becomes impossible to drive on it. There are important implications if buildings are on top of it'.

(Electronic Telegraph, 9 October 1997).

The position of the UK and Northern Europe might be even worse, for the possibility exists that while the planet as a whole gets warmer, the circulation system of the Atlantic might be disrupted. At present the warm surface water of the North Atlantic Drift, an extension of the Gulf Stream, is propelled northwards by a cold current, the Atlantic Conveyor, which is formed when cold salty water from the Arctic sinks and flows southwards along the bed of the ocean. If the Arctic Ocean were warmer, or if it were diluted and became less salty as the result of the melting of the ice cap, this effect might cease. The British Isles would then face temperatures found today in Spitzbergen, while Spitzbergen itself would be covered by ice. Furthermore, there is evidence that such periods of cooling have occurred in the past, without the additional stress that human activity places on the climate (The *Guardian,* 28 November 1997).

The implication of this is that in the event of continued global warming, the North may well be afflicted by different problems from the South, which would not only diminish its concern for the South's problems, but ensure that it was sufficiently preoccupied with other matters not to be able to help. In a worst-case scenario, the main grain growing areas of the United States, Europe and Russia might become too hot and dry to be able to act as a world reserve supply of food. For the present, however, the main problem is that these different scenarios, although not necessarily inconsistent with one another, create the impression in the North that the whole scientific basis for predictions of climate change is uncertain and possibly wrong.

GLOBAL WARMING AND ECONOMIC GROWTH

It is true that natural sources still contribute the greatest proportion of greenhouse gases, and this masks and seems to legitimize the increasing human contribution. The most important human impact on the atmosphere is the emission of carbon dioxide, mainly as a consequence of the burning of fossil fuels.

Some scientists, social scientists and other environmental commentators are now suggesting that global warming is caused by our whole way of life, and may call into question the wisdom of economic growth itself. As is often the case, there are arguments on both sides. Those who argue for point out that:

1. The AICs produce most CO_2 at present, and enjoy the privileges of having already developed. The USA alone accounts for 23% of CO_2 emissions, and President Clinton has recently set back the target for a fairly modest reduction to 1990 levels to 2012. The North's idea of development is not sustainable in the medium term, and is in any case incompatible with the aspirations of the South.
2. The South is set to follow. The massive increase in energy consumed between 1950 and 1985 was owing to increased economic activity in the South as well as in the North. As already noted, it is in the South that fossil fuel consumption has risen fastest. The South will not be denied what the North already has.
3. The world is already overheating and, given that emissions are continuing to rise, action is already almost too late.

However, there are also arguments against:

1. Natural sources still contribute the greatest proportion of greenhouse gases to the atmosphere.
2. Growth is essential if the ever-increasing population of the world is to be fed, clothed and housed; arresting emissions will slow growth in the South, imperil world stability and make it more difficult to act.
3. The temperature changes are real but fall within the normal range of fluctuation experienced over the past 10,000 years or so.
4. Others argue that a technological 'fix' has always been possible in the past and will always be possible in the future, though if it is, it may well be at such a high cost that it would be easier and cheaper to reduce emissions while it is still possible to avert the worst consequences of their release.

GLOBAL WARMING AND SEA LEVEL

In the present geological era the world has passed through several 'ice ages' or glaciations – though for much of what is now the South, of course, these were only periods of relatively cool, dry climate. During the last glaciation, which ended only some 12,000 years ago, the average temperature was only 4 °C lower than it is today. It is now believed that an increase of between 2.5 and 5.5 °C on the present figure would raise sea levels initially only as much as is due to the expansion of water at higher temperatures. Although the temperature increase from global warming is expected to be greatest at the poles, the vast ice sheets of Antarctica will take a long time to melt overall. While some sections of the ice sheets might melt in the shorter term, it is thought that others might thicken in compensation. Of course, sections of the ice sheet may break away altogether, as is already happening in the Weddell Sea. But floating ice melting does not cause a change in sea level; it is the thermal expansion of ocean water and the melting of land ice that will cause the sea level to rise.

As things stand, a sea level rise of only about 20 centimetres is expected to take place by 2030. However, floods and storms can be confidently expected to make matters worse. The Maldives, to take an extreme case, have no land above 2 metres above sea level. It has been estimated that a 1-metre rise in sea level would cost the Maldives US$10,000 per person in sea defences. Even their situation, however, could be better than that of the populations of the world's great river deltas in the South, which are home to and rich food production areas for millions of people. Large numbers of the world's poorest people will not have the means to defend their land against even small rises. Some 95% of Bangladesh is already at risk from flooding, for example. Monsoon shifts may be catastrophic for Bangladesh, as for some other tropical areas, although it is unlikely to experience much temperature increase.

The Pacific Ocean has the highest ratio of water to land globally (10,000:1), and four-fifths of the world's islands. It is therefore among the areas most vulnerable to sea level rises, and it is experiencing the same temperature increases as are being recorded elsewhere. The sea level in the Pacific has probably been rising for a couple of hundred years and is now rising at the rate of about 1.5 millimetres per annum. If the Intergovernmental Panel on Climate Change (IPCC) is correct, the acceleration of this rise will be devastating. Kiribati, the Marshall Islands, Tokelau and Tuvalu, for example, all lack any land over 4 metres above sea level. The largest Pacific islands are fortunate to have some high land too, and although 90% of their peoples live and work in the coastal areas, this leads to a perception among their power brokers that sea level rises are not their problem. And even if it is, like politicians everywhere they will wait to see how things evolve before committing scarce resources. In any case, it is not seen as a problem caused by them, so the rich countries which caused it should pick up the tab. But indignation, however righteous, will not alter the awkward fact that most commercial activity, such as sugar cane production in Fiji, takes place on low-lying coastal areas. Tourist attractions, especially the best beaches, are being lost too. Some aid is being given for shoreline protection, but there is an obvious reluctance to continue providing the help needed (Nunn, 1997).

> But the AICs will not be exempt from the consequences of climate change, either: in Massachusetts, a relative sea-level rise of 1.57 feet would submerge 10,000 acres of land at roughly $1,000,000 per acre. In New York, a three-foot rise during a major storm could flood Wall Street, along the Hudson and East Rivers and Battery Park, endangering the underground subway system. A three-foot rise in Washington, DC would allow the Potomac to flood a large portion of the Capitol Mall area during only a 50 year event. A recent study in California concluded that a two foot sea-level rise in Ventura County by 2040 could swamp hotels, power plants, the Point Mugu military base, roads, railroads, recreational facilities and as many as 4,100 low-lying houses. In Louisiana, the state's coastal wetlands are vanishing at a rate of about 35 square miles each year due to relative sea-level rise.
>
> (Leatherman, 1995).

CARBON EMISSIONS

Carbon emissions from fossil fuel burning reached a new peak in 1995 at 6056 million tons, compared to 5172 million tons in 1980 and 4006 million tons in 1970. Industrial nations are responsible for nearly half these carbon emissions, despite having only 15% of the world's population.

One hundred and fifty-seven governments signed the Climate Convention at Rio. Of these, one country, the USA, with 4% of the world's population, generates no less than 23% of global carbon dioxide emissions. Other major contributors are the former USSR (19%) and Europe (15%). The European Union has called for emissions to be not only limited at the 1990 level but reduced by 15% by the year 2010. However, the response of the USA has been particularly disappointing, and it is not just political neurosis that has led some Southern countries to feel that they are being required to conserve their rainforest to allow the rich countries to go on burning fossil fuels.

Only six countries agreed in 1992 to cuts in their emissions of carbon dioxide. The largest was Germany's target of a reduction of 25–30% by 2005, which would be partly met by phasing out inefficient plant in the east. For the United States, similar cuts in carbon dioxide would mean cuts in GNP of the order of 3% per annum – less than the military budget. However, the US administration did not agree to sign the Framework Convention until it had succeeded in weakening it to the point of meaninglessness, and the result has been that it imposes no real restraint on anyone. The Berlin Conference of 1995 showed that even the relatively weak targets established unilaterally by the major states had not been achieved, except where (as in the case of the UK) economic recession had fortuitously cut back emissions.

The Berlin Conference, however, was only one in a series of talks between 150 countries over two years, leading up to the Kyoto Conference of 1997. In Geneva in June 1996, the USA demanded a commitment from other states that all targets should be binding and legally enforceable, even though the promise made by President Bush at Rio in 1992 to hold emissions at 1990 levels had not been kept. At the review conference in New York in June 1997, President Clinton had promised: 'We will bring to the Kyoto conference a strong American commitment to realistic and binding limits that will significantly reduce our emissions of greenhouse gases.' Then, on the very eve of the Kyoto meeting, the newly appointed US Under-secretary of State for Global Affairs, Stuart Eizenstat, signalled an abrupt change in US policy. The USA, he said, was not prepared 'to have an agreement at any cost' and would not sign any new agreement to limit its emissions unless China, India, South Korea, South Africa and Mexico, among other developing countries, were prepared to sign too. Their failure to do so, he argued, would place the USA at a comparative trade disadvantage, something that hardly seemed to be as important as the future of the planet, but would undoubtedly get President Clinton out of an embarrassing confrontation with a reactionary Congress (The *Guardian*, 29 November 1997). A British Member of the European Parliament commented:

> Nowhere else are the problems of using a 19th-century political system with 20th-century morals to face the challenges of the 21st century more clear than in the US. A very small group of threatened fossil fuel companies are investing in lobbying to safeguard their short-term cash flow, at the expense of everyone else's long-term life flow. The tactics are well known: buy advice, buy scientists, buy time: claim to speak for all business and, if people don't like your arguments, change them to play on whatever fear does resonate.
>
> (Tom Spencer, The *Guardian*, 3 December 1997).

Even the most extreme reductions proposed for Kyoto, the EU targets, if met in practice would result on current assumptions only in reducing the rise in world temperature by 2050 from 1.4 °C (on the 'no action assumption') to 1.2 °C. This action, climatologists calculated, would save 'only' 20,000,000 lives directly, though it might additionally indirectly protect the livelihoods of many hundreds of millions. An agreement might still be worth having if it is followed by further real reductions and interim measures to combat the consequences of global warming (The *Guardian*, 2 December 1997).

In the event, the draft treaty presented to the Conference at its final session in Kyoto showed all the signs of a fudge. There was a target to reduce emissions worldwide by some 6% by 2012. But there was no specific commitment by the developing countries, who would be allowed to count carbon sinks (see below) against their emissions in calculating the quota. And the USA succeeded in gaining agreement that a country (i.e. the USA) could buy carbon quotas from other states that were already below target (such as Russia) in order to meet its own. These two concessions meant that even if the quotas were formally met there would be at best only a minimal reduction in emissions, and the carbon level would continue to rise but this was precisely the position agreed at the Buenos Aires Conference in November 1998 (see Chapter 8).

CARBON SINKS

Unfortunately, even the 'no action' assumption may prove too optimistic if there is no action at the same time to conserve the tropical rainforest, which, as the 'haze' of 1997 in Indonesia, Singapore, Malaysia and Thailand showed, is not only being destroyed but in the process contributing massively to the warming of the planet. Ecologists have argued that tropical rainforests, like the deep sea itself, should be regarded as 'global commons': something that the whole of mankind has an interest in preserving (*The Ecologist*, 1993). There is probably no single issue that more clearly illustrates the conflict of perception between North and South.

All forests act as 'carbon sinks', returning to the earth the carbon dioxide liberated by the burning of fossil fuels, and so acting naturally to arrest global warming. Their destruction, on the other hand, releases significant additional volumes of carbon dioxide to the atmosphere. In addition, their sustainable management as a rich habitat for a diversity of species could in time prove to be of immense and increasing value to the rising populations of the developing states. Tropical rainforest is particularly active. It recycles carbon dioxide quickly and generates a significant fraction of the world's oxygen (almost all the rest comes from the sea). Hence, rainforest is seen by scientists as a particularly vital 'carbon sink', helping to recycle surplus carbon dioxide and so avert global warming.

That is not at all how rainforest has been seen by local Southern politicians, who have been unable or unwilling to act to prevent its hasty destruction for short-term gain or ground clearance. In the early 1980s, it was estimated that some 27–40 million hectares of tropical rainforest were being lost annually; by the late 1980s, this figure had doubled. In the 1990s, some 15,000 square kilometres of Brazilian rainforest were being destroyed each year. After a disastrous year in 1996, the rate of burning actually accelerated in 1997.

The world's particular interest in the deforestation of the Amazon Basin stems from the fact that it is the last significant area of land surface to act to counter global warming. But for successive Brazilian governments the Amazon Basin has been seen as underdeveloped territory that they can use to alleviate the pressures of a growing population without having to carry out land reform. Even more crucially, settlement in Amazonia is seen as a necessary prerequisite to developing its resources for the good of Brazil and the fulfilment of its future destiny as a Great Power (see Chapter 7).

Under President Cardoso, the Ministry of the Environment, controlled by a conservative politician from the right-wing PFL party, has devoted the largest share of its budget to the drought-stricken North-East, switching money away from the protection of Amazonia. Hence when fires raged out of control in the state of Roraima in March 1998, there was a long delay while the state governor decided whether or not to fund firefighters and the response of the federal government was weak and indecisive (see also the discussion of the impact of forest burning on the biosphere in Chapter 6).

Again it is the way in which environmental problems interact which is illustrated by the diminishing of the rainforest in both Brazil and South-East Asia. Just as the loss of the forest as a sink is thought to have contributed to climate change, so aspects, some of them unexpected, of climate change will accelerate forest loss.

Such rates of change in climate are rapid by any estimate. A 1 degree C change is equivalent in the modern world in the mid-latitudes to 60–100 miles of latitudinal change. The significance of such high rates of change, never previously experienced in the modern world and never with 6 billion people on the earth, is profound, to say the least. With warming in the range of tenths of a degree per decade and the highest warming rates in the higher latitudes, the changes in climatic zones will be of the order of miles per year. The effects are uncertain, to be sure, but the probability is high that major changes in the climatic zones globally will result in warmer temperatures that speed the decay of organic matter in soils. Forests, their soils, and tundra soils of the Northern Hemisphere contain sufficient

carbon to add significantly to the annual emissions, thereby speeding the
accumulation of carbon dioxide in the atmosphere. These direct effects of
climatic changes will be amplified by the expansion of the ranges of insect
pests of forests, diseases of trees, and the increased frequency of fires
already observed. Such a positive feedback has not been incorporated into
current estimates of the warming. It is one of several potential surprises
lurking in the wings as warming proceeds.

(Woodwell, in Epstein *et al.*, 1996).

OZONE DEPLETION

Ironically, low-level *ozone* (O_3), which comes mainly from the reaction
of sunlight and pollution from motor vehicles, is a pollutant: a
powerful oxidant which causes damage to lungs and airways.
However, some tolerance develops, so smog may well be a less
important risk than workplace exposure: major sources industrially are
photocopiers, arc welders, X-ray machines and electrical equipment
generally.

On the other hand, the very small number of high-level ozone
molecules, in the outer atmosphere, form an essential shield against
ionizing radiation from outer space, and are therefore crucial to our
survival. Unfortunately, this shield is imperilled by chemical action
from two sources: chlorofluorocarbons (CFCs), formerly used as
propellants in aerosol cans and still used as refrigerants; and nitrous
oxide from artificial fertilizers. Worse still, low-level ozone will not rise
into the stratosphere to make up the deficit. Ozone is too reactive.

The biggest shock so far to world public opinion on the question of
climate change came from the discovery of the 'ozone hole' over
Antarctica. This discovery led in 1987 to the conclusion of the
Montreal Protocol to the Vienna Convention on the Protection of the
Ozone Layer, with the aim of regulating the use and release of ozone-
depleting substances such as CFCs and halons. The belated realization
that ozone depletion was accelerating in the Antarctic, and that it was
also taking place, though at a slower rate, in the Arctic, where it posed
a much greater threat to major concentrations of the world's
population, led to urgent measures to extend the Montreal Protocol.
The problem of the ozone layer was not directly addressed at Rio,
although it was very much a matter of concern to delegates from the
Southern Hemisphere.

Chlorofluorocarbons ($CFCl_3$ and CF_2Cl_2 are the most common) and
halons (similar compounds of carbon with chlorine, fluorine or

bromine) have been synthesized by human endeavour. Although they were 'thoroughly' tested when discovered in the 1930s, no one dreamt then that they would contribute disproportionately to the earth's warming, let alone that they would have damaging effects on the upper atmosphere. CFCs are used principally as the circulating medium in refrigerators and refrigeration plants. Although other agents are known (e.g. ammonia, butane, propane), they all have serious disadvantages, being too corrosive, too inflammable or too toxic. Halons are very stable compounds which are almost exclusively used in fire-fighting. Unfortunately, the very stability which makes them so effective means that they are very long-lasting and even small quantities in the atmosphere have the capacity to combine and recombine with very large numbers of ozone molecules.

CFCs are also alarmingly effective greenhouse gases. Molecule for molecule, CFC 11 and CFC 12 are 12,000–16,000 times more effective than carbon dioxide at trapping solar energy. Molecules remain in the atmosphere for some 60–90 years, so even tiny quantities have the potential to accelerate global warming. At the present rate, CFCs could contribute as much as 30% to global warming by 2015.

Production of CFCs is now below the level in 1965 and continues to drop following international agreements to phase them out. However, the recovery period even from this position is estimated at between 20 and 50 years, too long for any remedial action after first notification of the problem. Ozone-depleting substances already in the atmosphere will continue to be active for several decades to come, and stratospheric ozone is decreasing at the rate of 3% per decade: 3% depletion is believed to equate to 800,000 avoidable cancer deaths and 40 million cases of skin cancer.

Worse still, there are problems which suggest that this reduction in production may not be sustainable. CFC production in the South is increasing and accounting for a bigger and bigger proportion of the global output. In addition, a global blackmarket in CFCs is developing (Russia, India and China are producing; the USA and the EU are buying). Moreover, there is the impact on global climate change to be considered, as CFCs are also 'greenhouse gases'. Releases of CFCs (from old refrigerators, venting from industrial plants, use at fires etc.) could be of the order of 2.5–5.1% of current level per annum until 2025 and 2.1% per annum until 2050, which would raise the current level of 0.6 parts per billion (p.p.b.) to some 6–8 p.p.b., representing a thermal energy gain of some 70% over carbon dioxide emissions alone.

AIR POLLUTION AND ACID RAIN

Much air pollution comes from the lesser products of combustion. It is important to distinguish between *primary pollution*, caused directly by, for example, the smoke of traditional coal fires, and *secondary pollution*, typically caused by the photochemical action of light on the waste products of the automobile.

Sulphur dioxide (SO_2) comes mainly from the combustion of sulphur in oil and, more particularly, coal. As noted in Chapter 2, much of the coal currently burnt has a high sulphur content. In 1968, all natural sources of sulphur dioxide (microbes, volcanoes, sulphur springs, volatilization of sulphur as hydrogen sulphide (H_2S) by land plants and seaweeds, sea-spray, weathering of gypsum and other rocks) amounted to only one and a half times the total of human emissions, 90% of which arose in Europe, North America, India and the Far East. Emissions were highest in the 1970s, before controls were imposed in the AICs, but are now rising again as Southern states increase industrial production.

On its own, in small quantities, it is not very clear whether sulphur dioxide affects plants. However, it is strongly irritant to animal body tissues, especially the eyes and lungs. The warm, bright conditions of the Northern summer favour the formation of hydroxyl radicals (OH), which react with ozone and monatomic oxygen (O); exceptional levels can cause lung damage. Sulphur dioxide dissolves in rain water to form sulphurous acid (HSO_3), which, in the presence of traces of metal such as are generated by car exhausts and chimneys, is catalysed to sulphuric acid (H_2SO_4). The result is the acidification of soil, lakes and rivers, with potentially serious long-term consequences for the survival of existing species and the ability to accept new ones.

There are several different *oxides of nitrogen*, and we need to distinguish between them.

Nitrous oxide (N_2O, dinitrogen monoxide) is a natural product of biological processes in human and animal waste, but the pre-industrial concentration of 288 p.p.b. had increased to about 310 p.p.b. in 1990 and still increases annually by 0.25% (Elsom, 1992, p. 152). The main sources are motor vehicles, soil disturbance, the overuse of chemical fertilizers and biomass burning. In the presence of strong sunlight near ground level, nitrous oxide forms low-level ozone. However, it is also a greenhouse gas in its own right.

Nitric oxide (NO) is formed when fuel is burnt at high temperatures, oxidizing a small fraction of the nitrogen present in the atmosphere. The main sources of nitric oxide are power stations and motor vehicles, which between them account for more than 90% of all produced. Being relatively stable, it can rise into the stratosphere, where it depletes high-level ozone, and so helps to reduce the ozone layer, but in the presence of sunlight it forms the raw material for a series of complex chemical reactions which give rise to a range of secondary pollutants.

Nitrogen dioxide (NO_2) is a reddish-brown gas, formed in the atmosphere from the oxidation of nitric oxide in strong sunlight. It is the main visible component of photochemical smog, the characteristic blue-brown haze that hangs over most Southern cities. Other complex compounds formed include various aldehydes and some more complex molecules, notably the eye irritant peroxyacetylnitrate (CH_3COO_2-NO_2, commonly referred to as PAN).

Some 600 *unsaturated hydrocarbons* are produced by incomplete combustion of fossil fuels. The main requirements are the presence in the atmosphere of roughly equal quantities of oxides of nitrogen and hydrocarbons (acetylene, benzene, butanes, hexanes, pentanes, propane and toluene). Polycyclic aromatic hydrocarbons (PAHs) are known carcinogens.

Visible pollution from diesel engines and oil furnaces includes substantial quantities of *particulates*, very small solid or liquid-suspended droplets. Inhaled, they impair the efficiency of the lungs and may be poisonous (lead, cadmium), harmful (quartz, mica, china clay, asbestos) or carcinogenic. Reducing particulate emissions is the area in which the greatest progress has been made in the AICs and the least in the South. However, there is increasing concern that getting rid of visible pollution in this way may actually make matters worse, since the microscopic particles remain, and may indeed be increased by more efficient combustion processes designed to get rid of visible pollution. It is these microscopic particles which, because they are highly reactive, pose the greatest danger to public health.

The early oil industry was a major polluter and, as noted in Chapter 2, the process of locating, recovering and transporting oil continues to result in periodic environmental disasters. However, the biggest environmental disaster involving combustion of hydrocarbons to date occurred in the aftermath of the Gulf War, and was deliberate. The retreating Iraqi army set fire to 640 Kuwaiti oil wells in February 1991. Some four million barrels of oil, containing 2.5% sulphur, were burned off every day for several weeks, at only 70–90% efficiency, producing thousands of tonnes of oxides of nitrogen, 20,000–40,000 tonnes of sulphur dioxide and 100,000 tonnes of black smoke (particulates). Close to the wells, large quantities of unburnt oil fell to the ground; slightly further away, particulates and PAHs. Black rain/snow was recorded up to 2600 kilometres away in south-east Russia, Pakistan and Jammu and Kashmir in India; increased soot concentrations were recorded in Japan, Hawaii, Alaska and even the continental USA. Owing to the prevailing winds, very little fell on Iraq, illustrating the problem of transboundary origins, which makes the international politics of the environment quite different from other issues in international politics. It took nine months for 27 teams from 10 countries to extinguish all 600 wells (Elsom, 1992, pp. 192–5).

The problem of 'acid rain' is more insidious. Technically, it is known as 'wet deposition' of pollutants, but the popular term is more evocative. It was invented in Manchester, in the UK, as early as 1872. 'Rain-out' is an effective mechanism for removing pollutants from the atmosphere: dissolved bisulphite ions react to produce sulphuric acid, and oxides of nitrogen to produce nitric acid (HNO_3) ten times more rapidly, but other acids are also produced.

Deposition can now be monitored over a wide area. Acidity is highest on the Eastern seaboard of the USA and over a vast area of Northern Europe centred on Denmark. Significant increases in acidity after 1965 brought Europe-wide action to check emissions. The effects on soils vary, but in acid soils reactions can actually increase acidity further. It is important to remember (a) that rainwater is naturally acidic and (b) that there are natural processes of acidification, so even a small contribution from human causes can make a big difference (Regens and Rycroft, 1988).

'Acidification of freshwater is most marked in upland areas with high rainfall (hence high acid flux), steep slopes (resulting is a short residence time for water in the soil) and crystalline rocks, which weather ... slowly' (Andrews, et al., 1996, p. 97). Effects can be dramatic. Between pH 5.7 and 5.1 many freshwater animals are eliminated – the main effects of acid rain have therefore been seen on river and lake fish. Aluminium is toxic to fish, so the presence of even small quantities of the metal make the effects much worse (Wellburn, 1988). It is probably significant that humans biochemistry does not make use of aluminium. High levels of aluminium and low levels of calcium are both harmful to humans, and have been linked to Alzheimer's and other senile dementias, though the mechanisms are not clear.

With plants the situation is not so clear. Acid leaching impoverishes soils. Leaf injury cannot be unambiguously attributed to acid rain, but that does not mean that it is unimportant. It is also important to distinguish reversible from irreversible damage. Up to a point plants can absorb 'invisible injury', but it is difficult to assess to what degree, as it varies between individual plants and there is no sharp distinction between reversible and irreversible damage (Wellburn, 1988, p. 14).

In addition, the cost of atmospheric corrosion to materials of all kinds, ranging from water pipes to books, is estimated in the AICs to be between US$2 and 10 per head per year, a figure that is set to rise steeply in the South, where the marble facings of the Taj Mahal, for example, are already suffering serious surface erosion. So there are important economic savings to be made by controlling acid rain. Unfortunately, this is not as simple in practice as in theory (McCormick, 1989, p. 36).

The countries most subject to the problems caused by acid rain are not the same countries emitting the gases which cause the problem. Acid rain is not received in proportion to a state's contribution to it. Nor is it equally damaging in all situations. It is the countries with the most acidic soils, in Northern Europe rather than Southern Europe, which experience greater problems. Thus those contributing most are not so concerned about the problem, and those hardest hit do not see why they should pay or could not do enough towards the problem anyway. There are some situations in which funds from one country are being used to replant trees in another in recognition of the universal importance of carbon sinks destroyed by acid rain, but these are rare and rest on degrees of international cooperation that may be vital for our global future but are in short supply.

As early as 1972, the Stockholm Conference recognized acid deposition as an international problem, and the issue was taken up by the OECD. However, acidification is a growing problem for the South too, and the public health problem is compounded by heavy pollution, as the situation in China illustrates. There, all cities are heavily polluted by the practice of burning poor-quality coal. Some 60% of state enterprises experience no controls; in addition, the indoor burning of poor-quality coal is accompanied by very high lung cancer figures, especially among women (Mumford et al., 1987).

The main causes of air pollution in Southern states are much the same as in the AICs:

- the diesel engine (buses, lorries etc.);
- the internal combustion engine (cars, motorbikes, jeepneys, tuktuks);
- coal-burning factories and power stations (electricity is only 'clean' in the sense that dirt is concentrated in one place);
- fuelwood and open fires in towns, burning of vegetation in the countryside.

However, the main difference is in degree rather than in kind. Controls are minimal and a fifth of the world's population breathes air more polluted than World Health Organization (WHO) standards tolerate.

In Southern countries there is, inevitably, a strong incentive both to buy old designs of vehicles for their relative simplicity and to keep old vehicles on roads as long as possible: India's Ambassador, Sri Lanka's Morris Oxford and Brazil's VW Beetle are all ancient designs but still on the road. Cuba is a remarkable museum of old American cars: 1950s vehicles can be seen on the roads in every state from near-wreck to lovingly tended antique, and any vehicle that still goes after decades of the US embargo has some very strange spare parts inside it.

The same problems afflict all kind of industrial plant. Factories lack proper burners, so that excessive emissions are regarded as routine. Governments will not pay for clean power stations; they prefer to patch up existing plant or to go for hydroelectric power (which has its own problems).

> In Shanghai, China, 146 days a year on average exceed the WHO guidelines for SO_2, 104 days in Tehran, 87 days in Seoul and 68 days in Beijing. For suspended particulate matter, Calcutta experiences 268 days on average when levels exceed the WHO guidelines and Delhi 294 days. Beijing experiences 272 days per year with elevated particulate loads. Carbon monoxide concentrations in 15 cities studied by the UN Environment Programme (UNEP) in 1985 exceeded WHO guidelines.
>
> (Brimblecombe and Nicholas, in O'Riordan, 1995, p. 287)

CHEMICAL AND RADIOLOGICAL POLLUTION

Gaseous products of industrial processes are also released from dumps or emitted from oil refineries, petrochemical plants or factories, where controls are inadequate. Serious episodes of poisoning can result from simple failures in everyday processes. Twenty-three people died in Poza Rica, Mexico, when a faulty valve on a gas field installation leaked hydrogen sulphide for only 20 minutes (Wellburn, 1988, p. 185). If the escaping gas is inflammable, so much the worse. One of the worst industrial accidents of the century occurred in Cubatão, in the state of São Paulo, Brazil, on 25 February 1984, when petrol leaking from a pipeline under a shanty town exploded, killing 508 people. In the same year, a natural gas plant blew up in the middle of a crowded urban environment in the southern suburbs of Mexico City, killing 452 (Bogard, 1989, p. 41).

The most dangerous (and most common) hazard is *carbon monoxide,* which has poisoned people as diverse as French restaurateurs and Japanese silkworm cultivators. It is the most common atmospheric pollutant, equivalent in quantity to all others put together. Fortunately, soil acts as a 'sink' for carbon monoxide, having the capacity to absorb it to an extent that is believed to be well in excess of any possible need so far. It is as well that it does, since the oceans (other than surface marine creatures) do not absorb carbon monoxide in the way that they do its non-toxic relative carbon dioxide.

In Bhopal, the capital of the state of Madhya Pradesh, India, Union Carbide India Limited, a subsidiary of the US corporation Union Carbide, established a chemical plant in 1969, attracting substantial government inducements to do so. By 1984, the plant had expanded into a full-scale pesticide manufacturing operation in a facility covering some 80 acres. The plant, close to the railway station, was surrounded by both existing shanty towns and a wider urban sprawl, which had not been accompanied by any order to relocate. The result was a large chemical plant in the middle of one of India's poorest and most densely populated areas. Early in the morning of 3 December 1984, a violent chemical reaction in a storage tank released some 40 tons of methyl isocyanate (MIC), a lethal gas, which being heavier than air flowed through the surrounding slums at ground level. Hospitals gave up trying to count the dying, and after the event government agencies were unwilling to add to the recorded number of dead and injured, so it is now impossible to know how many were afflicted, but the best estimate available is that at least 2500 (between 2000 and 10,000) died, 17,000 were permanently disabled and between 200,000 and 300,000 were exposed in some degree, suffering various degrees of injury. The impact on animal and plant life in the city was equally catastrophic (Bowonder *et al.*,1985; Bogard 1989, p. viii).

Bhopal reflects the power relations of the developed and developing worlds, which affect not just the atmosphere but all aspects of the environment. Developing countries welcome investment and transnational companies seem to be prepared to tolerate (if not to encourage) practices that they would not risk at home. MIC is a highly reactive chemical, so dangerous that research chemists have been unwilling to work on it. It is capable of violent exothermic reaction in the presence of minimal quantities of contaminants. In AICs it is often not used at all for this reason, but, if it is, it is manufactured only as required and not stored in quantity. Yet on the night of the tragedy workers ignored the evidence of the notoriously faulty gauges and went off to their tea break instead. They returned to find the reaction already beyond control and the gas escaping. Investigators afterwards learnt that the refrigeration unit to the storage tank had not been working for some time, and none of the safety devices which should have helped to contain or neutralize an escape of gas were operative (Weir, 1987, pp. 40–3).

The Mexican and Brazilian disasters remind us that the power relations within Southern states also have to be considered. Local inhabitants seldom feel able to challenge the power and authority of a large state corporation such as Pemex, or to give up the chance of a relatively well paid job in a place like Cubatão, simply because it might be dangerous. They already know the risks of remaining poor.

CONCLUSION

An unprecedented research project, the Intergovernmental Panel on Climate Change (IPCC) recently released its consensus reports, stating that there is 'a discernible human influence on climate change'. The IPCC is a collaborative effort of more than 2,000 scientists from governments, universities, research institutes and the private sector world-wide, working on all aspects of climate change. Despite this remarkable research project, much of the public awareness regarding climate change has centered around whether or not there is even a threat from climate change. This distorted debate is fueled by a handful of skeptics relying more on the amplitude of their statements (and often industry provided platforms) than the clarity and originality of their research.

(Epstein *et al.*, 1997).

It is all too clear that in the years to come the Southern states will contribute far more than they have in the past to global warming. Pollution, on the other hand, is already a major problem, and the failure to deal with it is a clear sign that it will be even more difficult to enforce restraint on carbon emissions.

Airborne pollution offers one of the clearest examples of the strong economic pressures to continue to use old-fashioned and inappropriate technology. In order to compete, factories devote what capital they have to keeping existing plant going – modernization costs money and the many small companies involved wish to use existing investment to the full. The pay-back for the local community in terms of jobs outweighs any grievance they feel at living in a polluted atmosphere.

'It should be remembered that to some extent pollution is economically beneficial to some people. When a factory discharges waste into the atmosphere, it is presumably adopting the cheapest way of disposing of its unwanted materials. The act of polluting the air therefore keeps the price of its products lower than would otherwise be the case' (Elsom, 1992, p. 5). Consumers get cheaper products; factory workers get jobs. However others suffer the consequences but do not get the benefits; consequences may not even be felt in the same country or by the same generation. And cure is no substitute for prevention.

In fact, because the problem has almost invariably arisen slowly, over a period of years, residents hardly notice their worsening conditions and are inclined to accept them as part of the nature of things. Local politicians, dependent as they are on votes and money, show little enthusiasm for enforcing what environmental legislation may exist, still less for passing any more. The position may be different in the case of the TNCs, some of which have deliberately chosen to locate their polluting industries in the South to avoid environmental controls. However, protesters are inclined to be seen as troublemakers, the police to support the industrialists and the government to be all too aware of its limited capacity to change matters. In any case, it is the visible pollution of exhaust fumes and particulates that concentrates their attention, if it does so at all, and it is very hard to get reliable information about the extent of acidification in Southern countries, although logically it must already be well advanced.

The position is made worse by the demand for development, which concentrates populations into urban areas without affording them adequate support. Inefficient fuel use in poor homes results in the inadequate combustion which generates pollutants, with all the implications this has for public health. In-migrants resort for as long as they can to the fuel they know best, wood, thus accelerating deforestation. Although bottled gas is rapidly gaining popularity in Southern states, many of their inhabitants are simply too poor to switch to a less polluting fuel.

CHAPTER FOUR

WATER

Four-fifths of the Earth's surface consists of water. The oceans constitute the greater part of the vast reservoir that is the hydrosphere, and they are both the source and the sink for the hydrological cycle. That is to say, at any one time, substantial quantities of water are held in suspension in the atmosphere, ready to return to the surface as precipitation in the form of rain, hail or snow. This process is responsible for the production of the freshwater on which land-based life depends.

Freshwater makes up less than 3% of the world's water. Most of this (77.5%) is frozen into the polar icecaps and glaciers. Less than one tenth of the water that is evaporated, roughly 41,000 cubic kilometres, falls on land; of this only a third, or some 14,000 cubic kilometres, falls in the right places in a way that makes it available for human use. Although this is still a very large quantity of water indeed, and in theory more than enough to take care of human needs for the foreseeable future, in practice it is very unevenly distributed.

Moreover, although the technology for purifying it and ensuring it is fit to drink is simple enough, it does require an infrastructure. Northern states have been able to solve their problems. But the majority of Southern populations live in rural areas where only some 15% have access to clean water. Even in urban areas of the South, most households do not have running water. In India, for example, 70% of all surface water is polluted. The *World Development Report 1992* (World Bank, 1992) estimates that globally one billion (a thousand million) people are without access to clean water and 1.7 billion do not have proper sanitation. The combination of inadequate water supply and no sanitation is a guaranteed recipe for the rapid spread of water-borne illnesses. The World Health Organization says that the number of water taps per thousand population is a better indicator of health than the number of hospital beds.

Freshwater is becoming a more prized commodity as usage increases. Usage has tripled since 1950. Most is still used very inefficiently in agriculture, and much is lost to evaporation. Large-scale interbasin transfers are not the answer; they merely move the problem from one place to another, and at very high cost. Alternatives exist: in a report on a proposed project in Mexico in the early 1970s, a combination of smaller-scale works, improved irrigation techniques, local water savings and the possible use of injection wells was recommended (Cummings, 1974, pp. 108–9). However, Southern states seldom have the resources to undertake such works on their own. The United Nations Environment programme (UNEP) reported in early 1990 that a combination of drought and desertification had reduced the Lake Chad basin in area from 20,000 to 10,000 square kilometres and in volume by 60% over 30 years, but that a commission established in 1964 by the border states (Chad, Cameroon, Nigeria, Niger) had made little progress (*Keesing's Record of World Events* (hereafter *Keesing's*), 38259).

Not only is freshwater scarce, it is getting scarcer. The present-day figures are sobering. Global water usage doubled in the 40 years from 1940 to 1980 and is expected to have done so again before end of the twentieth century. In some parts of the world, water is already so scarce as to constrain both population growth and economic development. But, on the other hand, water has been one of the great successes of the twentieth century. The same figures can be regarded in a positive way, and indicate that in 1975 only 40% of people in the developing world had access to clean, safe water; by 1995 this figure had risen to 68%.

WATER AND AGRICULTURE

Domestic use is only a tiny part of human water use. Although it is vital for drinking and food-processing, freshwater also has to serve many other purposes, from washing away sewage to cooling industrial plant. Of the 14,000 cubic kilometres or so of renewable freshwater resources available each year, somewhere between 3250 and 4500 cubic kilometres are withdrawn for human use. Worldwide, 65–70% of this is used for agricultural purposes, almost 25% for industry and less than 10% for domestic use. In Africa, Asia and South America, agriculture is the primary use, with Asia using 86% of its water for agriculture, mainly for irrigation. Water withdrawals have been increasing by 4–8% per annum, with most of that increase in the South.

However, with a world divided in the way that it is, existing imbalances between states are very marked. China, with 25% of the world's population, has access to only 5% of the world's water. So many people need access to water that 35–40% of the world population lives in multinational river basins. Fifty countries have more than three-quarters of their territory in such basins. This has many important political consequences. Within a river basin system, there is a great advantage to the countries that lie upstream (the 'upstream riparian'). Not only can they count on more abundant supplies of clean water, but it is frequently not the polluter who bears the cost of contaminating water courses, as in the case of US pollution of the Colorado River, which reduces its value for irrigation downstream in Mexico. In 1997, the USA took so much water out of the river that it ran dry.

The main problem is the increased salinity of water when it has been used for irrigation. Other major problems created by human beings include contamination by the excessive use in agriculture of fertilizers and pesticides (less than 1% of all pesticides used are thought ever to reach a pest), by biological waste from human beings themselves and by industrial waste, particularly from large cities, but also from mines and other installations.

'Agribusiness', in increasing the scale of agriculture, is increasing the pollution of water. For example, pesticide run-off from the extended banana plantations of Costa Rica is threatening the Costa Rican eco-tourist industry. Trade liberalization will only intensify plantation growing of bananas, as Lomé Convention protection for small-scale producers in marginal areas, especially the Caribbean, is dismantled. Honduras, Ecuador and Costa Rica, along with TNCs like Del Monte, Chiquita and Dole, have already welcomed the outcome of the Uruguay Round of the General Agreement on Tariffs and Trade (GATT), and taken advantage of the new World Trade Organization (WTO) structure to try to drive out their small Caribbean island competitors (The *Guardian*, 27 June 1998).

Growing population, increasing urbanization (which lowers quality through sanitation problems as well as increasing demand), rapidly rising demands from industry and the increasing pollution of water courses by both solid and liquid wastes combine to make potable water a rarer and therefore more valuable resource. Water is also required for a variety of purposes other than drinking: washing and cleaning, irrigation for crops and, in more recent years, the generation of hydroelectric power.

Cooperation between countries in case of scarcity is vital. The most obvious case is the Persian Gulf, where the question of adequate water has from the beginning been the major issue confronting the Gulf Cooperation Council (GCC). 'The GCC has taken stock of the water balances and has found that water is in fact surplus to its needs. The problem, however, is one of distribution: you have to move it from places where it exists in surplus to places where it is in deficit. The whole area is sitting on a common aquifer. The states have a common reservoir, where the use by one country affects the use by another, and they really do need to co-ordinate this' (Kubursi, 1985, p. 68).

Currently it is confidently expected that water resources will become a major issue in international politics by the end of the twentieth century, when, it is argued, resource-based conflicts may provide new roles for underemployed armed forces. In one sense they have always done so. Water is so crucial to human life that access to it has always been a matter for the whole community. Like food, water can be used as a political weapon, with devastating effect. Elaborate irrigation works were constructed in ancient times in Mesopotamia, now Iraq. These canals and dams fell into decay in the time of the Romans, leaving desert in the upper reaches and marsh in the lower region between the Tigris and the Euphrates. In the past few years, Saddam Hussein has drained the marshes in order to destroy the habitat of his Shi'a opponents, the Marsh Arabs. The result of this is likely to be the further desertification of what was once the most fertile country in the Middle East, as the old term 'the Fertile Crescent' attests.

This use of water as a weapon has a long tradition, and in the Middle East there are several places in which international disputes continue over the use of water (Kliot, 1994). As the old cowboy adage has it, 'Whisky is for drinking, water is for fighting over.' It is a vital but vulnerable resource, and therefore one which enables some people to make others do what they would not otherwise do. In other words, it allows power to be exercised, and as such it is political. Today decisions about water tend to be made in bureaucratic organizations by technocrats and administrators, who then seek and get the support of politicians, always keen to be seen to be solving some problem or other in a conspicuous way. The most conspicuous way is to demonstrate control of rivers, the principal source of freshwater.

RIVERS

It would be hard indeed to overestimate the cultural importance of rivers. Since time immemorial, people have settled by rivers to enjoy the abundance of water for drinking and bathing. Travel by river has historically been much easier than travel by any other method. Rivers, therefore, are threads that join peoples and sustain life; hence, as Gita Mehta (1993) makes clear in *A River Sutra*, a source of both spirituality and wisdom.

Politicians and engineers, therefore, would be well advised to treat rivers with the respect accorded to them by local inhabitants. Unfortunately, all too often their view is that rivers are just large quantities of water which can be used to fulfil some very mundane needs. For them, often only the cost-effective or saleable aspects of river matter. For us all, rivers:

- provide freshwater for drinking;
- are a source of irrigation water;
- were and are major arteries for the spread and movement of human populations;
- form frontiers between states;
- provide essential supplies for industrial processing, cooling;
- keep up levels of lakes and inland waters such as Caspian, Aral Sea, Dead Sea (Hollis, 1978);
- form natural habitats for interdependent groups of animal and plant species.

Rivers can also constitute threats. Flooding has been a problem since the earliest times; yet it is now known that many of the traditional measures taken to control rivers have actually increased instability. Rivers carry away waste, but it has to go somewhere, and in the long run even the Mediterranean is threatened by pollution, salination and eutrophication (Myers, 1987, p. 93). The elimination of poisoning by industrial chemicals and agricultural run-off has had some success stories, such as, the cleaning up of the River Thames in the UK, but such examples are mainly found in Northern states that can afford the cost of such operations. In the South, and even in some of the transitional countries of Eastern Europe, this is not the case. In Poland, three-quarters of all river water is unfit even for industrial use (Swift, 1995).

RIVER BASINS

The land drained by a river and its tributaries is known as a watershed or river basin. As noted above, river basins contain a disproportionate number of the world's people, and important lessons can be drawn from each of the world's five largest river basins about the possibility or otherwise of sustainable river basin management (see Newson, 1992).

The Amazon

The Amazon Basin is dominated by Brazil, but shared with Bolivia, Peru, Colombia and Venezuela, and provides the most important hydrographical feature for the region, the bulk of which is under the control of Brazil. Significant hydrographical changes have already occurred in recent times as a result of increased human activity within the Amazon River Basin. Although the 6275 kilometre long Amazon River (Rio Amazonas) is not the world's longest river (that record is held by the Nile), its drainage basin of 4 million square kilometres makes it undoubtedly the world's greatest river. Many of the Amazon's tributaries are also of immense length: the Juruá River is 3280 kilometres long, the Madeira-Mamoré River 3240 and the Purus-Pauini 3210. The Amazon itself rises in the Andean lake of Lauricocha, close to the Pacific Ocean, and has over 1100 tributaries, of which ten carry a greater volume of water than the Mississippi River.

In view of the size and number of tributaries, it is perhaps not surprising that the outflow of water from the Amazon into the Atlantic Ocean is extraordinary. Every second the Amazon's 330 kilometre wide mouth discharges (depending on the season) between 34 and 121 million litres of freshwater and 35 tonnes of sediment into the Atlantic Ocean. The latter remains freshwater and coloured by sediment to a distance of between 160 and 325 kilometres from the mouth of the river. With regard to its use as a means of transport, the Amazon Basin is a veritable 'inland sea', since it provides over 80,000 kilometres of navigable waterways. Ocean-going vessels can sail as far as Iquitos in Peru, which is located 3885 kilometres from the river's mouth and is the furthest inland of any port serving ocean traffic.

The Amazon rainforest plays a key role in the regulation of the region's water cycle by ensuring an even distribution of rainfall throughout the region, leading to relatively stable river flows. Inevitably, deforestation disrupts the ability of the rainforest to carry out this function, and consequently presents the likelihood of significant climate change occurring across the region. Furthermore, since any exposed soil in the Amazon region is quickly compacted by the heavy tropical rains, water absorption is greatly reduced, and this then leads to soil leaching, ground erosion, river silting and widespread flooding. Deforestation of the Amazon rainforest therefore has very significant, if not irreversible, consequences for the region's water resources (Calvert and Reader, 1998).

Since the main stimulus to deforestation in the Amazon River Basin has come from the policy initiatives formulated by recent Brazilian governments (see Bunker, 1985; Goodman and Hall, 1990), it is the latter which must bear much of the blame for instigating the damage that has been suffered by the region's water resources. However, it is also necessary to point out that the main economic driver for the continuation of Brazil's unsustainable development policies in the Amazon region has been the country's need to service a gigantic and growing foreign debt (Cleary, 1991).

As already noted, a by-product of this development has been the pollution of the Amazon and its tributaries by inorganic substances, especially mercury from wild cat gold prospecting. Organic pollution, too, is a growing problem. Shortly after the outbreak of cholera in Peru in early 1991, the epidemic crossed the continental divide to become established in the headwaters of the Amazon. By the end of the year, the epidemic had spread downstream into Brazil, reaching as far as Manaus and Belém. Indian tribes living in the Amazon Basin are particularly vulnerable to cholera, since they drink untreated water taken directly from the rivers of the region, and from late 1992 onwards there have been reports of cholera-infected *garimpeiros* (prospectors) entering Indian territories in search of gold. (The *Guardian*, 20 April 1991, 19 December 1991; *Latin American Newsline*, January 1993).

The Nile

The Nile is the world's longest river (6670 kilometres). The White Nile flows from Lake Victoria through the giant marsh of the Sudd in the Sudan, before joining up with the Blue Nile at Khartoum. Rainfall on the highlands of Ethiopia used to make that area fertile, and created the regular annual rise and fall of the Nile and the 'inundation' which made Egypt a cradle of civilization. The Aswan Low Dam, built in 1902 to irrigate cotton fields for the Lancashire cotton industry, was furnished with sluices so as not to interrupt this natural cycle. The Aswan High Dam (1968) was not. The age-old rise and fall has been halted, bringing in its wake other major environmental consequences.

First, the river now loses one-fifth of its water to evaporation from Lake Nasser, the impoundment behind the Aswan High Dam. The scale of this loss can be appreciated when it is realized that it is equivalent to all the water annually consumed in the entire continent of Africa. The construction of the dam has had a dramatic effect on the river's ability to carry silt: only 8% of its natural suspended load is now carried downstream. (this is at the lowest end for any dam; the normal range is 8–50%, according to Goudie, 1986). Lack of silt has deprived fields of its fertilizing action and contributed to waterlogging and salination of downstream areas. Some 15% of Egyptian soil and one-third of the soil of the Nile Valley is now salinated, whereas the annual flood used to wash away the salt. The stagnation of the irrigation water has led to a dramatic increase in bilharzia or schistosomiasis (a debilitating and potentially fatal parasitic disease spread by snails), despite the fact that it is technically possible to reduce this risk substantially by a combination of good irrigation practices and hygiene measures as part of an integrated control package (Pike, 1987). Last, but not least, since there is no longer a plentiful supply of mud to make sun-dried mud bricks, builders have turned instead to quarried stone, with much more drastic consequences for the environment.

The certainty of the reserves and year-round irrigation enables the harvesting of three crops a year on the expanded agricultural land, but only with the heavy use of chemical fertilizers, the prices of which have become exorbitant as government subsidies have been removed. The price for this expansion of agricultural land is accelerated clear water erosion, both of the river bed, with effects on the watertable and drainage, and of much of the formerly highly fertile Delta, and with it the protection of coastal areas, which are now being destroyed at a rapid rate. At Rosetta, for example, the rate of erosion was 240 metres a year until expensive sea defences were built (El-Gawhary, 1995). Since the Aswan High Dam was filled in 1974 it has produced a 12 metre change in the water table. Interestingly, however, the overall area of irrigated land in Egypt remains almost exactly what it was in 1960.

Some of the consequences can also be partly attributed to 'channelization'. For some 1000 kilometres the Nile flows between embankments designed to restrain the unpredictable effects of the inundation. As the example of the Mississippi in the USA shows, building embankments can also have unwanted consequences: shifting flooding to areas downstream or upstream previously unaffected, increasing erosion and lowering the water table, or alternatively obstructing soil water movement and causing waterlogging.

The Congo (Zaire)

The Congo (about 4800 kilometres) is largely unaffected by problems yet, despite having the second largest flow after the Amazon. The River Congo forms the boundary between the Republic of the Congo (formerly Zaire) and Congo-Brazzaville — both have in the past few years been severely affected by political disorder and civil war, displacing tides of refugees. Deforestation has been limited and other developmental issues have yet to have much impact. Two large projects have already been built at Inga Falls, near the coastal outfall, but neither diverts the main flow and together they use only a small proportion of the available water. Plans have been discussed, however, between a number of states, including both Egypt and South Africa, to build a Grand Inga project which would exploit the full 40,000 MW potential of the Falls (McCully, 1996, p. 22).

Even more speculative plans exist to divert massive volumes of water from the Congo into the Sahel and, in the longer term, for the so-called 'Atlantropa Project' – a plan to build a dam across the Straits of Gibraltar, polder much of the Mediterranean Sea as land for neighbouring states and use the waters of the Congo to top up evaporation from the remaining Mediterranean Lake (Cathcart, 1983). Consequences for tropical flora and fauna in the semi-arid area are unknown, but can be guessed at. Indian experience suggests that dam construction may actually accelerate rather than solve the problems of desertification.

The Yangtze (Chang Jiang)

The Yangtse (5600 kilometres) is the longest river in Asia and is navigable by ocean-going vessels for some 2880 kilometres. All three of China's major rivers depend on channelization, as over the centuries the deposition of silt has raised the river bed. Rivers now flow above the level of the surrounding countryside for hundreds of kilometres. Failure to maintain the embankments, leading to periodic floods, was a major factor in the fall first of the Empire (1911) and then of the Nationalists (1949).

Dams, inevitably, have also been seen as a necessary part of the plan to tame the Yangtse. Gezhouba Dam on the Yangtse was originally planned to be opened in time for Mao Zedong's seventy-seventh birthday in 1970. Construction was planned to take five years, but immediately ran into technical problems and eventually took 18 (McCully, 1996, p. 19). Construction of a number of dams on the major tributaries of the Yangtse was accompanied in many cases by the forced resettlement of local inhabitants.

The latest and most ambitious plan is to dam the Yangtse at the celebrated Three Gorges to create the biggest dam ever built, a mile wide, with eight times the installed generating capacity of the Aswan High Dam. Work began in 1993. US, Canadian, Brazilian and British companies are all involved in the construction, the supply of materials and the financing of the project. The World Bank pulled out of funding in 1993, however, because of environmentalist opposition. Human rights issues remain. It will flood 350 miles of river valley and displace more than a million people. Apart from the irreplaceable loss in terms of scenery and culture, no one knows how the Three Gorges Dam will stand up to earthquakes. Failure of a dam on this scale would inundate much of central China.

The world's worst dam disaster so far occurred in Henan Province, central China, on 7 August 1975. A typhoon caused overload and failure of 62 dams of varying sizes: the largest was the Banquiao Dam on the Huai He, tributary of the Lower Yangtse, which burst when at least two metres over its maximum safe working capacity as a result of the sluices being partially blocked by silt; the collapse of the smaller Shimantan Dam followed. The most modest estimate from fragmentary statistics is that 85,000 were killed by the flood wave; 145,000 died from famine and disease afterwards. A Chinese source suggests that some two million were cut off by the flood lake for days or even weeks (McCully, 1996, pp. 115–16).

The Yellow River (Huang He), as its name suggests, is one of the world's muddiest rivers. Flowing for some 4000 kilometres through the loess soils of central China, it is so turbid as a result of erosion and run-off that the river bed is continually rising. Since the first major works on the river in 2536 BCE, it has been channelled and in places the river now runs some eight metres above the flood plain. Periodic flooding has resulted, with heavy loss of life (Wijkman and Timberlake, 1984, p. 60). Hence, the urge to dam it for flood control and at the same time harness it for hydroelectric power was irresistible.

Sanmenxia (Three Gates Gorge) Dam was constructed between 1967 and 1970, with technical assistance from the former Soviet Union. Some 410,000 people were displaced (so far a world record) and 66,000 hectares of prime farmland flooded. However, the impoundment silted up so badly within three years that floods upstream threatened the ancient city of Xi'an, and sedimentation caused serious damage to the turbines. To flush through the sediment, therefore, the engineers had to abandon altogether the dam's original proposed function of retaining river water and to restrict its generating capacity to a mere 250 MW, a very poor return indeed for so much environmental damage.

The Ganges (Ganga)

The Ganges (2400 kilometres) rises in the Himalayas and flows southward to form a vast delta on the Bay of Bengal. The problems of the Ganges, India's sacred river, include its intensive use for drinking, washing and ritual purposes. The Indian government has objected to use of its waters by Nepal before they cross the frontier, and there has been hot debate about the effects of deforestation in Nepal. The latest evidence, however, suggests that deforestation peaked in Nepal between 1890 and 1930 and that other factors are to blame. However, India itself has contributed to its problems by failing to control deforestation, much of it generated by the need for fuelwood and/or the need to house India's growing population (the value of fuelwood increased dramatically as a result of the 1973 oil crisis, accelerating pressure on dwindling stocks). However, large-scale flooding, such as killed 237 people in India in 1985, and the even more disastrous floods in Bangladesh in 1988 and 1998, could not have been prevented even if the original forest cover was intact (Mannion, 1991, p. 254; see also Brammer, 1990).

Earthquake-damage to dams threatens more than 100 villages and a substantial town. There is a great deal of water-borne pollution, mainly industrial but also due to cremations at Varanasi (Benares). Urbanization has been an important factor increasing flood-water runoff and the risk of flooding.

Plans for large-scale inter-basin transfer to the Narmada and other rivers seem at best speculative. However, a potential conflict arises over both the quality and the quantity of the water shared between India and Bangladesh.

DAMS

Dams have impressed politicians and public alike, and even now are hailed by some as a great achievement. Jawaharlal Nehru called them 'the temples of modern India' (Sachs, 1992, p. 231), enabling the conquest of the environment and the harnessing of natural power. The megadams or superdams of today, defined as those with a height greater than 150 metres to distinguish them from 'large' dams of more than 15 metres in height, are the largest structures ever built.

The movement to build large dams began in nineteenth-century Britain, but the capacity to construct them was limited by the available materials: earth, stone and mortar. In the twentieth century, reinforced concrete became available and the achievements of the USA (Tennessee Valley Authority, Hoover, Grand Coulee etc.) were proclaimed by Americans, spurred worldwide admiration of economic progress through the control of nature and were quickly imitated by newly independent states. Similar propaganda came from the USSR, where the Dneiprostroi Dam on the River Dnieper was completed in 1932, submerging large tracts of prime Ukrainian farm land. Control of the Volga and Don followed in the post-war period. Since the 1960s, the scene has shifted to the South, but expertise in building (as we shall see) remains with the North.

At best dams bring:

- regulation of river discharge and prevention of flooding;
- controlled power for electricity generation;
- water storage and irrigation to desert areas;
- employment in construction and industry.

For some countries they may be seen as having strategic or political importance, and as enhancing national prestige. But this can be at a high cost. The Aswan High Dam, for example, enhanced Egypt's standing among its Arab neighbours; it also made it potentially much more vulnerable to an Israeli attack.

A decline in the number of completions reflects the disappointment with the apparent benefits of large dams. Further, the benefits are very unevenly distributed and some (i.e. employment) are only temporary. The damage done, on the other hand, may take a long time to become apparent, and little of it is evident until the irrevocable decision has been taken and the machines start work. Negative consequences include:

- Triggering subsidence and even earthquake (reservoir-induced seismicity).
- Loss of river fishing.
- Loss of land rights by those dispossessed.
- Loss of wildlife habitat above and below the dam.
- Transmission of harmful organisms under changed land/water conditions.
- Loss of the river as a source of water, and its replacement by a very inefficient use, involving piping water long distances and evaporation from the standing water behind the impoundment.
- Reduction in sediment load downstream; the fertile flood plains of the 'mature' river are destroyed and silt lost; the need for fertilizers and irrigation leads to high costs, environmental damage and salinization. Sind Province in Pakistan was once productive, but now irrigation schemes resulting from the damming of the Indus have left more than half of it saline and barren.
- Clear water erosion downstream, changes in groundwater levels, the

need to rebuild/reconstruct waterworks and sewage outfalls, waterlogging of soils.

- Costs are borne by the poorest after the debts incurred by national governments lead to the imposition of SAPs; there are opportunity costs in poor nations, where the size of the schemes gives rise to escalations of costs and corruption.
- Recent evidence shows that even in the absence of mercury contamination, the creation of impoundments leads to dramatic increases in the leaching of mercury from the underlying soil and its concentration to unacceptably high levels in fish eaten by local inhabitants (McCully, 1996, pp. 39–40).

People in developing countries are least able to cope with some of these consequences of large dams. However, when considering the damaging effects, one should also keep in mind two key arguments for large dams:

1   About 3 billion people in the world do not have electricity. It would be very difficult to argue that they do not have a right to it. Civilization is rural electrification.
2   The world's food production must double over the next 40 years. There is no realistic way this can be achieved except by continuing to expand the wet cultivation of rice.

WHY DAMS?

Given the numerous problems, the continuing building of large dams may seem strange, but there are clearly pressures which encourage dam building. In the last analysis, it is the power structure that determines where and when dams are built. However, the local elites are not the sole group at work. McCully also identifies the key role in promoting dams of:

- Northern funding agencies. Dams look impressive and the agencies have been slow to question their real value. The World Bank has spent more on dams in the past 50 years than anything else, and its role has been vital in dam building. Since 1948, when its first grant was for the construction of three dams, it has invested US$50 billion in 500 large dams in 92 countries, which have displaced some 10 million people. There are now 36,000 large-scale dams and the number is increasing at the rate of 170 per year (Swift, 1995). Each year close to a million refugees are displaced, mainly the weak and vulnerable in the South. They do not get – and could not afford –

the expensive hydroelectricity. In 1986, the US Congress enacted a requirement that beneficiaries of dams met all the costs, and dam building was effectively stopped in the USA. The fact that the South is so dependent on external funding means that no such legislation will be passed there.

- Northern engineering consortia. Would-be dam-builders offer substantial inducements to local politicians (who, for reasons of status, may not need persuading anyway). The dam-builders (not just the engineering and construction companies but also the World Bank and other financial representatives) are members of the International Commission on Large Dams, which meets every three years. The overall effect is to promote further construction and to 'play down' the environmental damage.

- Northern consultants, who produce the environmental impact assessments (EIAs) demanded by funding agencies, when they mistake 'monitoring' for 'correction' of their harmful effects.

He concludes:

> Dams invariably cost more than claimed, diverting investments from more beneficial uses. Reservoirs tend to fill with silt long before predicted and hydroplants to supply much less electricity than promised. Irrigation schemes are badly managed, destroy soils, bankrupt small farmers and turn lands used to feed local people over to the production of crops for export. Dams assist the powerful and wealthy to enclose the common land, water and forests of the politically weak. By misleading people into thinking they can control huge floods, dams encourage settlement on floodplains, turning damaging floods into devastating ones.
>
> (McCully, 1996, p. 24).

Not surprisingly, there is often strong local resistance to dam projects. This has been overcome by a combination of bribery and intimidation. The rights of indigenous inhabitants have regularly been ignored in environmental assessments by consultants who take at face value government assurances that 'no one' lives in the affected area, or that people can be relocated on the higher, steeper slopes of valleys without cost to their lifestyles and cultures. In extreme cases, local inhabitants have been deported: in one year, 1958, 57,000 from the site of the Kariba Dam, Northern Rhodesia (now Zambia); 21,000 from the Miguel Alemán dam, River Papaloapan, Mexico. Or they may be massacred. Government forces slaughtered 378 Achi Indians in one day at Río Negro in March 1982, to make way for the Chixoy Dam in Guatemala. The World Bank gave more money for the project later, and does not mention the massacres in its 1991 completion report

(McCully, 1996, pp. 72–6). After removal, mortality among the displaced rises sharply; they typically lose much of their land and all of their common rights. New land 'given' to them is more marginal, is often barren and lacks basic facilities; otherwise it would already be occupied.

Grassroots resistance and NGO activity seriously worries construction interests now. Dam building in developed countries (where some talk of actually reversing previous dam building and draining the impoundments, as in Tasmania) is largely blocked now by environmental concerns, and the World Bank has been pressured out of several large projects. NGOs such as the International Rivers Network (established in 1985) target funders, especially the World Bank, and expend most of their efforts learning how the World Bank functions, who has influence there and what its current thought is. But it is important that they also involve local people in saying what their needs are, and this is much harder to achieve.

South Asia

The most successful campaign against dams thus far is the River Narmada campaign. The Narmada rises in the mountains of north-central India and flows for 1280 kilometres almost due west to the Gulf of Cambay. The multipurpose Bargi dam, situated on the Narmada in Madhya Pradesh, was completed in 1990 at the cost of 113,600 displaced persons. There was massive grassroots resistance. For example, 40,000 local people turned out in Harsud in 1989 to protest at the Sardar Sarovar project, also situated on the Narmada in Gujarat, which if completed would have displaced some 320,000 people. This project, along with the Narmada Sagar Dam in Madhya Pradesh, which would have displaced a further 300,000, is still stalled at present. Unfortunately, although the Indian government has decided to pull out of ten sites, it took over the Narmada Project when the World Bank developed reservations about its value, and is most reluctant to abandon it completely.

Other stalled projects in India include the Tehri Dam on the Bhagirathi, a tributary of the Ganges, in Uttar Pradesh. In 1992, the Indian government promised two-year suspension and a review of the scheme; in 1995, construction was resumed after police set on protesters, but it stalled again soon after. In the meantime, there is water rationing in Delhi, the capital of India. There they naturally want both the water and the power that a dam such as the Tehri Dam could provide. Since it is in Delhi that national decisions are taken, it is hardly surprising that they are pressing strongly for the project to go ahead (*Assignment*, 'Lands of the Dammed', BBC Television, 1993).

There are good reasons why the opposition should be taken seriously, for while the loss of land to impoundments is real enough, the benefits are invariably overstated and often illusory. Not only do irrigation projects consistently fail to fulfil their promises, but the few who do benefit are rich landowners and agricultural corporations, not the farmers who have been displaced. Not the least of the reasons for opposition to the Narmada project is the evidence that is already to hand that very few are likely to benefit, while many are going to lose out. 'One of the biggest irrigation fiascos in India is Bargi Dam on the Narmada, which submerged nearly 81,000 hectares of farmland and forest to irrigate a projected area of 440,000 hectares. Although the dam was completed in 1986, seven years later only 12,000 hectares were receiving irrigation water (3 per cent of the planned area)' (McCully, 1996, p. 167; see also Vajpeyi, 1994).

But concern about the effects of the megadams goes much further. Water in the wrong place can have serious consequences, and the sheer size of the dam projects has raised the question of the stability of the subsoil under the weight of the huge volumes of impounded water. Studies have been undertaken to ascertain the stability of dam structures in the face of possible seismic activity, but until recently seismologists were reluctant to accept that the dams themselves might trigger earthquakes. Over the past 20 years, however, the frequency of earthquakes in the Narmada Valley seems to have increased. In 1995, Dr Arun Singh recommended that the seismicity of the Narmada Valley be re-evaluated. On 22 May 1997, a strong earthquake of magnitude 6.0 killed some 50 people and injured more than a thousand in Jabalpur, Madhya Pradesh. Its epicentre is believed to have been between 20 and 40 kilometres from the site of the Bargi Dam (McCully, 1997).

Organization of resistance in India has been primarily local, using Gandhian methods, such as refusing to leave homes and having to be forcibly removed, organizing rallies and hunger strikes. But protest groups have also employed legal stalling tactics, up to and including appeals to the Supreme Court. The main issue in India, however, is not the damage to the environment but the displacement of the local inhabitants, which is seen not in its environmental context, but simply as an abuse of land rights. More than 16 million Indians have already been displaced, and each state has its dam conflicts. There are other good reasons for opposition, though. The Tehri Dam is supposed to be 250 metres high, the biggest in India, if completed, but it is located only a few kilometres from the epicentre of a major earthquake that occurred in 1991. Factionalism within the Indian anti-dam movement may have prevented a national movement developing, but 'the model for successful action to save rivers from megadams remains passionate local activism backed up by the sophisticated media-manipulation of Western Green groups' (Pearce, 1995, p. 30).

Africa

Power relations also govern decisions in Africa. In Nigeria, for example, where power rests with the military government in Abuja and local power with Kano, capital of Northern Nigeria, not with the villagers and nomadic cattle-herders of the Hadejia-Nguru wetlands abutting the fringes of the Sahara Desert. The wetlands are dying because upstream water is being diverted by the Nigerian government to supply Kano and to irrigate the fields served by the government-run irrigation scheme, which are usually owned by those with political or military connections. Three dams are already built and the largest yet, at Kafin Zaki, is being undertaken by a German company. The water table is dropping dramatically in the wetlands. For every one hectare irrigated by the diverted waters two wetland hectares dry out. Fisheries are being lost. The dams are promoting the desertification that their proponents claim they will help to arrest.

East Asia

The World Bank has refused to fund the Three Gorges Dam on the Yangtze, just as it in 1995 withdrew funding for the Arun III Project in Nepal. But the result is privatization; a process of international capital giving way to national capital has led on to a return to funding through international capital.

There has been some displacement of dam building into neighbouring states. Most of Laos lies within the watershed of the Mekong. Traditionally villagers have made small dams to irrigate their paddy fields. Some of the 7000 irrigation systems are hundreds of years old. Massive dam building in Laos is taking place to supply Thai industry, since, in Thailand, some dams have been stopped as a result of popular unrest. Electricity is now Laos's third biggest export earner. The result is that foreign companies are now logging virgin forest in Laos to clear it for more dams. Dams and forest clearance are meanwhile causing the breakdown of traditional irrigation systems and the destruction of traditional fishing. In Thailand the construction of houses, golf courses, etc. on the edge of the new lakes brings fresh problems.

Americas

For many decades Brazil's North-East has been the site of Brazil's worst poverty. The principal cause for this has been the existence of what is known in Brazil as the 'Drought Polygon'. This is an area greater in size than Western Europe, encompassing six of the North-East's nine states, that has been subjected to more than 70 *secas* (droughts) since the Portuguese colonists began keeping records of these catastrophes in 1587. The special problems of the Drought Polygon have long been recognized by the Brazilian government. However, despite the establishment in 1959 of a special government development agency for the region, known as SUDENE (Superintendência do Desenvolvimento do Nordeste), most of the measures taken by the government to improve the situation have until now had little effect. As with most of Brazil's welfare-related problems, the failure of considerable financial investments to provide a long-lasting cure to the difficulties of the Drought Polygon can be attributed to the combination of three factors: a 'big is beautiful' approach to development that fails to take proper account of the needs of the majority of the population; a lack of administrative and organizational competence, which fosters (and is fostered by) corruption; and an extremely high level of social inequality, which ensures that political and economic power is concentrated in the hands of Brazil's wealthy elites. Brazil has the most unequal wealth distribution in the world, and the North-East has the most unequal wealth distribution of Brazil's regions.

Many of the North-East's water development projects have amounted to little more than political gestures, and most of those water reservoirs that are eventually completed are underused. Much of the money allocated for these projects is misappropriated by members of the North-East's governing elites. The latter ensure that the various construction contracts go to their relatives or friends, who manage to make sizeable profits from pharaonic dam building schemes that produce few tangible benefits for the general population. According to a newspaper report published in 1993, a secret World Bank study found that only 40 dollars out of every 1000 going into the region as aid actually reach the drought victims. This situation is unlikely to change while funding bodies channel their aid through the region's elites, who have grown rich and powerful as a result of Brazil's 'industry of drought' (Susan Branford, The *Guardian*, 20 February 1993,

p. 9). By controlling the flow of aid to their voters they assure themselves of their support at election times.

The São Francisco River, informally called Velho Chico and popularly known as the 'River of National Unity', is the longest river wholly within the borders of Brazil. The river rises in the Central Brazilian Highlands and travels 1609 kilometres northward until it turns east into the Atlantic Ocean. Because of the presence of the huge Paulo Afonso Falls, where the São Francisco River cuts through the 'Great Escarpment', only the last 277 kilometres of the river are navigable by ocean-going ships. The hydroelectric potential of the river was initially tapped at the Paulo Afonso Falls, but the cyclical droughts that limit the river's usefulness for transportation gave rise to a project to build a new dam at Sobradinho. The history of the Sobradinho Dam project is illustrative of the problems associated with the North-East's 'big is beautiful' drought relief schemes. Originally intended for irrigation purposes, the dam was instead used to provide hydroelectric power for the industrial centres of the region. Building commenced in 1973, and the dam was inaugurated in May 1978, after having displaced some 72,000 people and submerged dozens of towns and farms under one of the largest artificial lakes in the world.

The completion of the Sobradinho Dam made the livelihoods of the small farmers who lived along the banks of the São Francisco less rather than more secure, since they could no longer count on the natural flow of river water and their lands experienced a range of new problems, ranging from salination to heavy flooding. These problems forced many of the small farmers into bankruptcy, and as a result they had to sell their lands to the region's wealthy farmers, thereby increasing further the concentration of land-holding in large farms (Chilcote, 1990). More recently, in early 1996, President Fernando Henrique Cardoso announced his government's intention to invest US$500 million in the development of a 2000 kilometre canal system, which would use the waters of the São Francisco to provide 6 million people in the dry North-East with access to piped water. However, given the history of water development schemes in the North-East, there must be scepticism over whether it will actually improve the living standards of the mass of people in Brazil's poorest region (*Brazil Network*, May/June 1996).

Drought, however, is not the only reason why Brazil has chosen to build megadams, and some of the environmental consequences have been even more alarming. Currently most of the power needed for the Carajás iron ore project (Projeto Grande Carajás, PGC) is provided by the massive Tucuruí hydroelectric dam on the Tocantins River in

North-East Brazil, which will produce 7700 MW when all its turbines are completed. The dam's construction necessitated the displacement of 20,000 people and destroyed 2400 square kilometres of rainforest when it was filled with water in 1984. Prior to the filling of the dam, much of the forest cover was removed by the same chemical defoliant as was used during the Vietnam War. Rotting vegetation in the completed reservoir produced noxious fumes, killed the fish and bred swarms of mosquitoes, which made life unbearable for villagers within a 30 kilometre radius of the dam. However less than a fifth was removed, and the massive piles of rotting vegetation remained, deoxygenating the river water and generating huge quantities of methane (McCully, 1996, pp. 38–9).

Another dam associated with the PGC is the Balbina Dam on the Uatuma River, 160 kilometres north of the city of Manaus. The dam was completed in 1988 at a cost of over US$800 million dollars, but is unlikely to produce a fraction of its relatively low planned output of 250 MW, since the reservoir suffers from high evaporation rates and produces a very low flow rate at the turbines. The dam displaced 6000 Amerindians (many of whom were shot or poisoned by those who wanted the dam built quickly), while the flooding of the dam's 1454 square kilometre reservoir destroyed a huge area of rainforest, only some 2% of which had previously been cleared, and interrupted the natural hydrological cycle of the entire river basin (Pearce, 1992, pp. 212–16).

Further dams planned for the eastern part of the Amazon under *Plano 2010* include the Altamira–Xingu complex, which consists of a series of six dams located on the Xingu River, upstream from the city of Altamira on the Transamazon Highway. The prospect of the Altamira–Xingu complex creating the world's largest reservoir led to it becoming a focus of international concern, especially when it was realized that it would lead to the displacement of about 70,000 indigenous tribespeople, a significant rise in water-borne diseases and the loss of over 7000 square kilometres of rainforest. In 1989, joint protest actions by tribespeople, *seringueiros* (rubber-tappers) and various international environmental groups culminated in the highly publicized Altamira meeting, which obliged the World Bank to reconsider its decision to fund the project (Cummings, 1990, pp. 63–75). The state of Matto Grosso in Western Brazil is the scene of the latest, but possibly the last, of the great dam projects, Porto Primavera, which is the focus of considerable hostility among conservationists.

It takes so long to plan and bring a dam project to fruition that mammoth cost overruns are normal and estimates of their economic utility are consistently wrong. Such is the case of Itaipú, on the Paraná River where it forms the border between Brazil and Paraguay. From the beginning it was known that Paraguay could not consume all its share of the power generated; instead, it hoped to sell it at preferential rates to Brazil to fuel industrial development in the southern Brazilian states of Santa Caterina and Rio Grande do Sul. Economic recession in the later 1980s, however, resulted instead in a marked drop in demand, so the additional power was not needed. Likewise, in the case of the Yaciretá Dam, downstream on the Paraná from Itaipú, on the border between Brazil and Argentina, when the first turbine started operating in 1994, Argentinian demand had similarly failed to keep pace with the rosy projections of the dam builders, and the country already enjoyed a substantial surplus of installed generating capacity.

Dams are often represented as being non-polluting, and hence sustainable. However, it is important to take into account all factors in making this sort of calculation. First, impoundments generate large quantities of methane, a much more efficient 'greenhouse gas' than carbon dioxide; hence, wherever calculations have been possible, a hydroelectric project can be shown to have many times more impact on global warming than a fossil fuel plant of the same output. The Balbina Dam on the River Uatumá in Brazil, for example, generates 26 times more greenhouse gases than a comparable coal-fired plant (McCully, 1996, pp. 140–5). Moreover, because of sedimentation, all dams also have a finite life. This can be prolonged by flushing through water, and this is done, but at the cost of reduced electricity production. No one has yet worked out the costs of decommissioning megadams and scooping out the huge quantities of sediment that accumulate behind them, since at present this is not a realistic proposition. And very few large dams provide significant quantities of drinking water – they are simply too far away from the places where water is needed for this to be practicable, and in lowering the water table downstream large dams have actually made the problem of securing safe drinking water more acute, a serious cost to a Southern state and its population.

SMALLER DAMS

More modest projects seem to have struck a better balance between the benefits they offer and the problems they generate.

The Victoria Dam on the Mahaweli River in Sri Lanka was partly funded by the British taxpayer. It was planned in 1978 and built in six

Table 4.1 Large dams and their problems

	For	Against
Water	Regulation of river discharge	River fishing destroyed Transmission of harmful organisms Water wasted by evaporation from standing water behind dams Leaching of mercury
Energy	Electricity generation	Debt, escalation of costs and corruption Costs incurred by roads, transmission lines, ground clearance
Land	Irrigation to desert areas	Inefficient use, e.g. evaporation from standing water behind dams Reduction in sediment load downstream Fertile flood plains of 'mature' river destroyed and silt loss Loss of land rights by those dispossessed Loss of wildlife habitat above and below dam
Economic	Employment in construction and industry	Subsidence and earthquake triggering 'Clear water' erosion downstream, changes in ground water levels Environmental damage and salinization Forced resettlement Tilts power relations towards the rich

Source: based on Patrick McCully, *Silenced Rivers: the Ecology and Politics of Large Dams*, London: Zed Books in association with The Ecologist and International Rivers Network, 1996.

years. The dam provides power for Sri Lanka's burgeoning textile industry, boosted to the level of a national economic ideology by the late President Premadasa. It also supplies irrigation from the wetter west of the island for the drier east. Rice production was increased fivefold using the dam's potential for irrigation. Sri Lanka used to import food that now it exports. The turbines of the Mahaweli project provide not only the power for textile production but also electricity for the towns of Sri Lanka; not the villages yet, but that is expected by the end of the twentieth century. Social considerations were given priority in the displacement process. Whole villages were moved together. Their biggest problem remains what it was before the move: marauding wild elephants.

On the negative side, the wild elephants have now been cut off from their traditional migration routes and hence constitute more of a problem than they did. The Premadasa government failed to observe ethnic balance in allocating land, and instead used the project as an opportunity to favour Sinhalese over Tamils. There were health problems. The first outbreak of malaria for years occurred in 1986–7 after the loss of wild animals led the mosquitoes breeding in stagnant pools to turn their attention to the settlers instead. Hepatitis was also a problem at first for those who were displaced and resettled, although most soon developed immunity. Environmental issues were not given the priority they would certainly be accorded now; they were simply not seen as so important. There has been a loss of biodiversity; the natural habitat of some of Sri Lanka's rich wildlife, including seven endangered and two threatened species, has been flooded. The irrigation canals have deteriorated more quickly than predicted. There is also some evidence that cutting back the forest in the region of the dam may have made the local climate hotter (Maya Institute, 1978; Alexis, 1983).

Megadams are very inefficient compared to small-scale river works, such as those in use in western China, which harness the power of the flow without impeding it. Megadams will become even less efficient over time as they silt up. Small dams (microdams) are sometimes cited as the sole 'solution'. However, it is clear that small dams could not begin to equal the hydroelectric power potential of large ones, and they are not free of environmental consequences. What can cope with the problems, however, is a combination of techniques:

- reducing demand for electricity;
- water conservation, dry farming, prohibition of wasteful uses of water on, for example, golf courses, lawns, in desert areas, more efficient use of rainwater;
- alternative energy sources, e.g. wind power.

THE OCEANS: GENERAL

The oceans cover four-fifths of the planet. They have several important roles.

First, the sea is the great motor of climate, a massive reservoir of energy: tides, currents and waves all have their influence on climate patterns. The differing temperatures of water in the tropical, temperate and polar regions drive the circulation of air between zones. But these patterns are regular rather than constant. For example, the periodic impact of El Niño (ENSO) in the Pacific, when the concentration of warm water normally to be found off New Guinea moves eastwards until stopped by the isthmus of Central America, in the process diverting the cold Humboldt Current, brings heavy rains to the Pacific coasts of the Americas, in the Caribbean quietens hurricane patterns but in the Indian Ocean results in a delay to the onset of the monsoon season.

Second, the sea is both the source and the sink for the hydrological cycle. Evaporation makes freshwater available to sustain all life on land. Every year the energy of the sun evaporates a quantity equivalent to more than a metre of all the seas and oceans in the world. The vast majority of this falls directly at sea. When it falls on land, rivers return used water (and any pollution) through their estuaries to the sea.

Third, the sea moderates the temperature of coastal regions and gives unique characteristics to those regions (including islands). This is specially important in the Southern hemisphere, which is mostly water. This has a great importance for the climate and conditions of the South in the geographical sense and some effect on North–South relations in the political sense.

It is easy to imagine, as many do, that the sea is too big to be affected by human action, and it is true that up to now the impact on the world's oceans has been relatively limited. However, in one respect it is clear that there is already global impact: global warming, with the threat of more violent storms and the ultimate melting of the polar ice caps.

COASTS

Half the global population lives in low-lying areas, within 60 kilometres of the coast. It is expected that by 2020 this figure will rise to 70%. Yet the factors that make the coast attractive also make it unstable and even dangerous.

Strong winds and tidal waves may threaten the coastal environment and its densely packed inhabitants, especially in the region of the trade winds. Further, if global warming brings, as is already expected, a rise of anything up to a metre in mean sea level over the next century, it will be the Southern states that will suffer most. The ten countries most vulnerable to sea level rise are, in order of vulnerability: Bangladesh, Egypt, The Gambia, Indonesia, The Maldives, Mozambique, Pakistan, Senegal, Surinam and Thailand (Ince, 1990, p. 59).

The sea makes ports possible. Ports are centres of trade and commerce, and very attractive to those in search of work. But in the past, coastal erosion, silting up and/or the effects of isostatic readjustment have rendered previously thriving ports useless. A sea level rise can have a much more rapid effect, especially where water abstraction has caused the land surface to sink, as is often the case in large cities. Major ports such as Calcutta, Lagos, Cairo, Rio de Janeiro and Bangkok all face the prospect of flooding. The land is subsiding under Bangkok, for example, at the rate of between 5 and 10 centimetres a year (Neale, 1996). But none of the countries concerned is likely to be able to meet the bill, even assuming it were technically possible to provide coastal defences. Shanghai, on the Yangtse Delta, presents especially complex problems, since China, with its massive, silt-laden rivers and long, winding coastline, is particularly vulnerable to sea level changes. 'A future rise in sea level of several centimetres will have serious impacts on China's coasts, especially on the low-lying coastal plains and deltas where there is rapid subsidence of the land at rates higher than those of the sea level rise' (Wang and Zhao, 1992). Shanghai is sinking both because of natural causes and because of the extraction of water for the thirsty city from underground aquifers, but its 186 kilometres of sea defences were planned and built on assumptions that are already out of date:

> One of the principal disasters to affect the coasts of China is storm, especially in the case of cities on large estuaries. Because of the interaction between run-off and tidal currents, a combination of typhoon and exceptional high astronomical tides can lead to serious disaster. Sea-level rise will aggravate the situation in the future. For example, Shanghai City ... has an elevation of 4 m according to the Wusong datum but the height of the average high tide for the majority of its history has been 5 m; so industry and agriculture as well as the security of the population of the area depends on protection by sea walls.
>
> (Wang and Zhao, 1992, pp. 167–8).

There are several other ways in which the environment of coastal regions is significant. Coasts contribute some 25% of total biological productivity. Over two-thirds of the world's fisheries and nine-tenths of shellfish production take place in coastal areas. Tides and currents around the coast have traditionally dispersed seeds; they have also aided navigation for human beings. Wind flow is affected by contact with land, and precipitation results, differentiating humid and dry zones. Saltwater marshes are staging grounds for migratory species on account of the abundance of food and relative safety of the environment. However, as noted above, dams drastically reduce the ability of rivers to carry silt. This affects not only the rivers themselves, but also estuaries, marshes and the many small sea creatures which live on them. Silt from the Nile used to create a big fishing industry off the delta by attracting fish to the feeding grounds; although fishing has recovered slightly in recent years with improved methods, yields are still well down on those known earlier in the century.

Development has upset patterns established over millennia, and has either direct or consequential effects on the environment. Above all, it means the pollution of the coastal environment by organic and inorganic waste, as, for example, in the case of the water surrounding the islands on which the port city of Bombay (Mumbai), India, was built. According to an official report published in January 1995:

> Sewage in the city is not treated before discharge into the Arabian Sea at any of the three Corporation areas. All sewers overflowed into coastal waters adjoining Bombay, which made them unfit for recreational use throughout the year. Hundreds of septic tanks overflow into the ground, causing flies and mosquitoes to breed. Two million people live with no toilet facility. Drinking water supplies have to be delivered by tankers to supplement piped water. Over 5.5 million people live in slums where enteric and respiratory disorders are common and gastroenteritis, tuberculosis, malaria and filaria are 'rampant'. Each day, despite the work of recycling, Bombay produces 5,000 tons of garbage.
>
> (Seabrook, 1996, pp. 45–6).

Pollution of coastal regions is compounded by a tendency to regard the problem as self-correcting. However, there is plenty of evidence to show that neglect is potentially lethal. Molluscs are sensitive to pollution and can be used as indicators of water quality. Very small quantities of diesel fuel (0.1 millilitres/litre) produce harmful effects and prolonged exposure 20% mortality. The effects are greatly increased with raised temperature (Tapp et al., 1996). Pollution of shallow coastal waters by biological effluent, apart from having disgusting effects for beaches and swimmers, creates a well known danger by contamination of shellfish (oysters, mussels, shrimp, crab, lobster). Biological contamination rapidly renders molluscs and crustacea unfit for human consumption.

Pollution by nitrate run-off from fields is another potential problem in the coastal zone, especially where it is enclosed by land or coral reef. Nitrates in limited quantities can be beneficial, stimulating the regular annual algal growth in spring and later summer, on which zooplankton feed. Sometimes, however, such blooms can become so large as to be clearly visible, typically where there is a downwelling in the sea to concentrate them during their daily vertical migration. These blooms are commonly known as 'red tides' because they can appear rust coloured, but the colour varies according to the species involved. In recent years they have been becoming much more frequent, probably because of increased run-off of fertilizers.

Red tides can be disastrous to other marine life in two different ways. If conditions change, the entire population can die simultaneously, their decomposition exhausting all the oxygen in the water and suffocating other animals. Moreover, not all algae are desirable; at least two dinoflagellate genera include species which produce toxins that can kill fish and even have serious effects on birds, marine mammals or humans that come into contact with them. Such toxic blooms are rightly regarded as a serious environmental hazard and are a regular occurrence in the Gulf of Mexico, off the west coast of Florida. Contamination of shellfish by toxic algae has long been known to be responsible for paralytic shellfish poisoning of people who consume them, and this too has been becoming more frequent in recent years (Waller, 1996, pp. 58–9).

The worst problems are faced by the small island states of the Pacific Ocean. All the islands together have a combined land area of only some 500,000 square kilometres. Essentially they are all coast and, not surprisingly, therefore, they depend overwhelmingly on the resources of the ocean and are extremely vulnerable to the conditions that for most other states would give rise only to local coastal effects. Given the normal vulnerability of small islands to flooding by hurricane and tsunami, and the small number of land species available to them, these islands represent at best, despite their considerable charm, a very marginal habitat for human beings. And because of their small size, the outside world may be very slow to notice that they have sustained a serious disaster, such as the devastation of Tonga by Hurricane Isaac in 1982.

In the long term, the main threat to the islands is more fundamental: the death of the coral reefs on which they depend. Corals are extremely sensitive to rising sea temperatures and Pacific surface temperatures, have now risen to 28 °C. Loss of the reefs would mean loss of protection for low-lying coastal areas from tropical storms, especially as sea level rises. Coastal development also harms coral in a number of ways. Most corals are very sensitive to the salt content of the water in which they live, so the discharge of freshwater can kill them, as may many other pollutants of human origin, such as organic waste or disinfectants. Mechanical damage by excavation, by the use of grappling irons or fishing tackle or even by tourist souvenir-hunting destroys these complex habitats much more quickly than they can repair themselves.

The local inhabitants, with so little land at their disposal, have long been used to dumping old cars and other things at sea. The long-term implications include the loss of biodiversity and of income from tourism, probably the only foreseeable additional economic resource for many island communities. Outmigration from the islands over the past few decades has been very heavy and, for example, today more of the nationals of Tokelau live in Wellington, New Zealand, than on their own island (Ince, 1990, p. 94).

There has, sadly, been a long history of the North regarding the South Pacific as so remote that it can safely be used as a venue for nuclear tests (Bikini Atoll, Christmas Island). It is fair to say that the carelessness of the 1950s reflected the then poor understanding of the cumulative dangers of radiation. However, this excuse is no longer available. The flagrant disregard by the AICs for the welfare of the local inhabitants should have come to an end with the series of nuclear tests conducted by France at Mururoa Atoll.

As land-based mines are worked out, mineral exploration is likely to turn increasingly to the sea around the Southern states. In Brazil, little petroleum has been found on land. The main Brazilian fields lie some 100 kilometres out from the country's north-east coast. Maritime production is a mainstay of both Argentina and Chile. However, mineral exploitation at sea carries with it substantial risks of pollution to the marine environment.

Additionally, most pollutants are poorly dispersed in coastal waters. Chemical pollution by heavy metals, pesticides (DDT, organochlorines) and polychlorinated biphenyls (PCBs) has been well documented and is concentrated in whales and dolphins, which, being placental mammals, can pass it on to their young (Waller, 1996, pp. 408). The main sources of most of these are specific industrial processes. For example, the main source of mercury is the chloralkali industry, and it has been reliably estimated that in each year in the 1980s between 85 and 90 tonnes was being discharged into the sea in a form which enabled it to be taken into the food chain. Cadmium, classified as a blacklist substance, is emitted by both industry and by municipal sewage plants. Pesticides containing aldrin, endrin and dieldrin, already widely accepted to be too toxic to be safe at any level, are still used extensively in the South.

OIL

Oil pollution kills cetaceans, but indirectly harms them by poisoning the environment, and/or kills many species of fish by entering the food chain. Since much of the world's oil is produced in Southern states and marketed in the AICs, the scope for marine pollution at all stages of that chain is considerable, and this raises important questions of legal liability and environmental protection.

Pollution of the seas by oil tankers is alarmingly frequent. Despite improved navigational aids and techniques, in the 1990s major marine oil spills have occurred with monotonous regularity, with the vast majority involving a tension between AIC and Southern interests. The most obviously preventable are marine collisions (Thompson, 1993). In January 1993, the Danish-owned *Maersk Navigator* collided with an empty tanker and spilt some 25,000 tonnes of light crude into the Malacca Straits, sending a slick some 50 kilometres in length drifting towards the Nicobar Islands (*Keesing's Record of World Events* 39298). In July 1995, 154 k.p.h. winds from typhoon Faye drove a Cypriot-registered, 140,000 tonne tanker on to the shore of Yeosoo, South Korea, contaminating a 50 kilometre area and causing considerable loss of marine life (*Ibid.* 40646). Such cases are of course less well known than those that have hurt the maritime environment of the AICs themselves. But the source of the problem is the same, the dominance of free market ideology propagated by the North, which not only drives the productive process but also has led to squalid and often unnecessarily dangerous conditions at sea and a virtual end to the sort of protection for sailors which Victorian legislators such as Samuel Plimsoll argued was essential. It is abundantly clear that the outflagging of tankers to Southern countries with weak regulatory regimes was and is an extremely irresponsible policy. The idea that Liberia, for example, a country which in the early 1990s was in a state of open civil war, was able to police conditions aboard its nominally vast fleet would be laughable if it were not so dangerous. The irony is that the poor sailors who die often come from Southern states such as Korea and the Philippines.

A series of highly publicized disasters at sea, however, tends to take the spotlight off the day-to-day actions of oil companies. Not all such companies are Northern transnationals: the state-centric legacy continues to exist alongside free market ideology, and it is not necessarily less polluting to the environment. In early 1996, there was an 18-day blockade by indigenous groups of 64 oil installations belonging to the Mexican national oil company Pemex in the State of

Tabasco, protesting (among other things) at inadequate compensation for past oil leaks over many years (*Keesing's* 40946). A major catastrophe in the Gulf of Mexico, already polluted by off-shore operations and from leaks from land-based operations on the Panuco River, could have been foreseen many decades ago. However, the Mexican government refused to accept any kind of responsibility when in 1979 the offshore well Ixtoc 1, drilled on behalf of the Mexican national oil company Pemex, sent a torrent of oil gushing northwards towards Texas over a period of several months. On this occasion the responsibility of the Mexican government might have been the greater, since it refused international help as matter of national pride, though well-capping was and is highly specialized work. In 1983, the US government received US$2 million in compensation from the owners of the drilling rig, in a settlement requiring a covenant from the USA not to proceed against either the drilling contractor or Pemex as the well owner (22 *International Legal Materials* 580 [1983], cited in Smith, 1988, p. 117).

In February 1991, the retreating Iraqis, as well as setting fire to Kuwaiti oil wells, deliberately pumped 6 million barrels of oil into the Persian Gulf, as discussed in Chapter 3 (Elsom, 1992, p. 195).

DUMPING

Seas have long been seen as a safe dumping ground for inorganic and organic waste. But with the AICs tightening up their laws on dumping, the focus of attention has turned to Southern states, where, all too often, the culprits are large Northern companies, or, at least, the underlying market imperatives, which encourage the cutting of corners by those they subcontract.

In 1994, lawyers for the French petrochemical giant Rhône-Poulenc sought an out-of-court settlement when contractors dumped 12,000 tonnes of waste contaminated by hexachlorobenzene (HCB) from a plant making pesticides and solvents in Cubatão at various points along the coast of Brazil between Rio de Janeiro and Santos. All but one of the 159 workers in the plant had been found to be contaminated when the plant was closed in 1993, and four had since died. Fish in the area of São Vicente were found to contain 4750 times the 'safe' level of HCB (The *Guardian*, 27 October 1994).

Bad publicity and increasing legal constraints are having some effect. In September 1992, two European companies that had planned to dispose of industrial and hospital waste by dumping in Somalia signed an agreement not to do so, following UN intervention (*Keesing's* 39123). In March 1994, the 64 signatory states of the 1989 Basel Convention on the Transboundary Movement of Hazardous Wastes agreed to ban all exports of hazardous wastes from the AICs to Southern states, with immediate effect in the case of material for dumping, and from 31 December 1997 in the case of material for recycling. Australia, Canada, Germany, Japan and the UK, however, all initially opposed a total ban, while the USA, which was not a signatory to the Convention, continued to oppose it (*Keesing's* 39939).

There are well documented cases of ships from AICs simply 'fly-tipping' cargoes of hazardous waste in Southern states or their coastal waters where no protest was likely to be made or officials could be bribed. The sea gives freedom of navigation: it is possible to move things a long way from their point of origin in the hope they will not be noticed. But the 'freedom of the seas', like the equality of opportunity principle underlying other aspects of the market ideology, means that the existing advantages enjoyed by the North enable Northern nations to exploit opportunities and 'freedoms' more readily.

> The freedom of the seas has been the prevailing principle. But the principle could be abused. When technology advances by leaps and bounds and makes it possible for both animal and mineral resources to be exploited, this prompts a reaction from countries which are not yet technically capable of such exploitation. It is a well-known fact that enormous fishing armadas from industrial states 'cleaned out' sea areas off the coasts of the developing countries in the quest for fish. It is also well-known that an advanced mining industry saw great opportunities of extracting minerals from progressively greater depths under the seas, far beyond the purely legal definitions of the term 'continental shelf'. Most of the Third World countries objected to this 'uneven distribution' of exploitation.
>
> (Theutenberg, 1984, pp. 2–3).

The 'Third World' countries, however, have been systematically outmanoeuvred. In 1945, President Harry S. Truman of the USA claimed for his country the right of jurisdiction over the so-called 'continental shelf'. In the 1950s, Latin American states followed by claiming a 200-mile territorial sea. The first Conference on the Law of the Sea, held at Geneva in 1958, accepted the 'continental shelf' principle, though both it and a second conference two years later, in the same place, failed to endorse the principle of actual sovereignty, moving instead towards the notion of an exclusive economic zone. Meanwhile, the idea of a 200-mile zone spread among the newly independent African and Asian states.

At the Third United Nations Conference on the Law of the Sea

(UNCLOS), delegates concluded in 1982 a new Law of the Sea Convention, which accepted the principle of broad zones for coastal states and actually extended them further by allowing states to mark out wider areas using as the base straight lines drawn from cape to cape. Hence, although the Convention in effect rejected the idea that the sea bed and ocean floor were, in the words of UN General Assembly Resolution 2749 (XXV) of 17 December 1970, 'the common heritage of mankind', it codified what was already established practice, rather than making new law, when it left only the ocean floor open to all (Theutenberg, 1984, p. 8).

The Convention was adopted not by consensus, as the text had been, but by 130 votes to four, the four being the USA, Israel, Turkey and Venezuela. Not surprisingly, therefore, the United States refused to ratify it, relying instead on 'creeping jurisdiction' to give it what it wanted without having to incur reciprocal obligations. For example, the Law of the Sea Conference had in effect declared open season on the rich nodules of manganese and other minerals that (for some reason as yet unexplained) litter the ocean floor. The Northern states reasoned that since they alone would have the resources to access them, the Convention provision was hardly necessary. Among the 17 abstentions were the UK, the German Federal Republic, Italy, the USSR and its East European allies; these last, however, subsequently acceded to the Convention when it was opened for signature in December 1982.

Part XI of the Convention lays on coastal states the duty and obligation to protect and preserve the marine environment. They are bound additionally by the *sic utere* principle: the obligation to use one's property so as not to harm one's neighbours. It is specifically the duty of coastal states to monitor and report on pollution and to take all necessary steps to control pollution. States have the right to investigate the actions of foreign-flagged vessels for this and other purposes, provided that they do it in a non-discriminatory way.

CONCLUSION

The Law of the Sea Convention came into effect in 1994. However, the tendencies already noted persist. The Northern states (led by President Clinton of the USA) have successfully put off action to restrain carbon emissions until 2012. In consequence, we can now expect that the rise in mean sea level, coupled with increased run-off, building on flood plain and land subsidence through water extraction, will be sufficient to bring coastal flooding and similar problems within the lifetime of most people now living. The slow warming of the oceans, which maintain the stability of our weather patterns, is also bound to continue. It is not yet certain whether the disintegration of the Antarctic ice shelf (which, though spectacular, has little effect on sea level) is the prelude to more drastic changes.

Meanwhile, a crisis in freshwater supplies may be nearer than we think. Things have gone seriously wrong with the well intentioned programme of sinking tube wells in Bangladesh to provide its inhabitants with clean drinking water. Although the water is, as was intended, still biologically safe, it has proved to be so seriously contaminated by naturally occurring arsenic that poisoning episodes have occurred and at present there is no obvious solution to the problem; an unusual variant of economic pressures to use inappropriate technology.

When there is competition for scarce resources, the water industry attracts little interest and the provision of sewage disposal even less. Former colonies, such as Cuba and Sri Lanka, are still using the infrastructure laid down in the colonial period for a much smaller population. The problem of the lack of capital to modernize facilities is also found in the AICs themselves.

The perceived costs of environmentally friendly solutions also result in the use of rivers and coastal waters for the unrestrained dumping of both liquid and solid waste. Individual poverty and poor living conditions cause organic pollution of both water and the creatures that live in it. There is, therefore, a close link between poverty and the spread of water-borne disease. The run-off of fertilizers is associated with toxic blooms and the overuse of pesticides with contamination of shellfish.

Last, but not least, the vogue for large dams exemplifies the distorting effects of dependence on foreign sources of investment. Large lending agencies, whether private or public, naturally tend to favour large projects, and inevitably governments seek to use the money they provide to maximize their political, rather than their environmental, advantages. The alarming thing is the way in which so often, in the governmental mind, the relatively small numbers of people who seem to stand in the way of these pharaonic projects are simply reclassified as uninformed and therefore unimportant.

CHAPTER 5

THE LAND

The theory of plate tectonics is a classic instance of how a scientific hypothesis, when and only when the evidence for it becomes overwhelming, becomes accepted, and old paradigms are abandoned. The continents are carried on underlying plates in constant motion over the liquid surface of the Earth's core, which has remained far hotter than earlier scientists predicted because of the then unknown phenomenon of radioactivity. Zones of instability occur at the edges of plates. These lead to: (a) the so-called 'European' band of volcanic activity, stretching from the Mediterranean to Indonesia; and (b) the so-called 'Pacific ring of fire' and its associated volcanoes and earthquakes. Most of the world's land-based volcanoes occur near the line where one plate passes under another (subduction); where this is accompanied by sideways motion of one plate against another, as with the San Andreas fault in California, USA, the Anatolian fault in Turkey or the Alpine fault in New Zealand, seismic activity is particularly evident. Although most of the rifts between plates are covered by sea, there are some 250 active volcanoes where the rifts are visible. Hawaii illustrates a third possibility: a chain of volcanoes produced by the movement of the earth's crust over a local 'hot spot'. Ninety-five per cent of the world's known volcanoes, however, occur along fault lines, and it is there that pressure by one plate on another creates mountain ranges and hills; the most recent major range being the Andes, created by the pressure of the edge of the Nazca plate against the South American plate.

There are believed to be many more thousands of volcanoes under water, of which some 20 are in eruption in any one year. In the long term, some of these may eventually build up into conical islands, an example being Loihi, a submarine mountain 30 kilometres south-east of the island of Hawaii, the top of which is still more than a kilometre under water. Once extinct, such islands in tropical zones are often fringed by coral reefs, protecting them from storm and wind and facilitating navigation by human beings.

GEOHAZARDS

The political problem is that, although geological instability affects both developed and developing countries, it does not do so equally (McCall *et al.*, 1992). Southern states are more vulnerable and political conditions there make the likelihood of a major disaster much more likely. It also makes it more difficult for them to cope with natural disasters, and the reporting of these disasters, in turn, tends to shape the public perception of the South in the AICs.

The case of volcanoes is illustrative. The largest share of the volcanoes that are known to have been active at some time in the past 10,000 years are in the USA or on US territory. Among the Southern states, the leading country for volcanic activity is Indonesia, with 127 volcanoes. The explosions of Tambora in 1815 and Krakatau in 1883 count among the world's largest natural catastrophes. The Philippines has 51 volcanoes, of which 21 are still active, second only to Indonesia. As recently as 1951, some 500 died in the explosion of Hibok-Hibok, on Caminguin Island, south of Mindinao, and in 1965 the eruption of Taal, near Manila, killed 235 and caused destruction within a radius of 50 kilometres (De los Reyes, 1992). In Mexico in the spring of 1982, El Chichón erupted for the first time in recorded history. More than a thousand people died and many more thousands were rendered homeless. The ejection of dust and sulphuric acid was so violent that it was believed to have contributed to a temporary fall in global temperature and a severe winter in 1982–3 (Decker, 1991, p. vii).

There are important lessons to be learnt, although the chances for most people of being near an erupting volcano are negligible. In fact, almost all the people killed so far by volcanoes in the twentieth century died in only two incidents, both in the South: the eruptions of Mont Pelée in Martinique in 1903 and Nevado del Ruiz in Colombia in 1985. The latter case shows how politics plays a major part in the impact of disaster. The site of the eruption was a long way from the national capital and the problem was not a priority. Both local and national politicians were reluctant to act; instead, they sent the scientists' plan of the danger zones place back for revision, although there was no evidence of inaccuracy, and the actual mud slide which overwhelmed the town of Armero occurred almost exactly as predicted. Scientific equipment to monitor the activity had to be borrowed and had been returned before the disaster actually took place. There was no one and nothing high on the mountain to give a warning when the mud slide began; fast as it was, the slide took some two hours to cover the 60 kilometres down the Lagunillas canyon to

Armero, and with adequate warning the 22,000 inhabitants who died could have been evacuated to safety (Hall, 1992). Hall concludes: 'The overall responsibility of the Ruiz crisis management and resulting tragedy lies with the national government. The factors that hindered the Government's role include its skepticism about an eruption, inadequate leadership, a non-responsive bureaucracy, and other distracting emergencies' (Hall, 1992, p. 52; see also Decker and Barbara, 1991, pp. 3–6).

As will be seen from the above, earthquake activity is associated with volcanic eruptions. But earthquakes are much more common, especially (but not exclusively) along the fault lines caused by continental drift. Japan suffers some 700 earth tremors a year. Unfortunately, although delicate listening equipment and some surface indications can give warning that an earthquake is likely to occur soon, it is impossible as yet to say just when it will occur, and in both Japan and California, large modern cities have been built along known fault lines despite all historical evidence. Nor is it possible to estimate with any accuracy just how severe the shock might be if an earthquake does come, although engineers in earthquake-prone areas build if they can to a substantial margin of safety. The Kobe earthquake of January 1995 was much more severe than anything that had been predicted for that part of Japan, measuring 7.2 on the Richter scale. More than 5000 died, and 310,000 were made homeless, and there was severe criticism of both the slowness of the relief effort and Japanese reluctance to accept international assistance. Despite the scale of the damage, the resources of a sophisticated AIC could be made available for both rescue and reconstruction much more quickly than they would have been in a less advantaged and/or more remote area (*Keesing's*, 40359).

Social deprivation and political disorganization contribute massively to the loss of life in earthquakes in the South. On 4 February 1976, a huge earthquake struck Guatemala City, capital of Guatemala in Central America, and ten other towns in the immediate area around. Severe damage occurred even to the city centre; loss of life was greatest, however, among the shanty towns perched uneasily over the ravines at the edge of the city, and the effects of the disaster were compounded by the disruption of all forms of transport, 400 kilometres of road having to be reconstructed (Plant, 1978).

For a Southern state, too, its relations with the outside world are of prime importance in determining whether a disaster is a crisis or a catastrophe. In 1973, after the Managua earthquake, Nicaragua received massive aid from the USA on account of the client relationship of the Somozas with that country; most of this aid was subsequently embezzled. In 1974 when neighbouring Honduras was struck by hurricane Fifi, the Hondurans got virtually nothing, since the US ambassador felt they did not need it. However, between 4000 and 8000 people were killed and hundreds of thousands made homeless. Ironically, it was the US-owned banana corporations that had driven them out of the fertile land of the valleys and up into the mountains, where the heavy rains turned the deforested hillsides into liquid mud (Wijkman and Timberlake, 1984, pp. 79–80).

In the Philippines, which has been shaken by more than 70 major earthquakes between 1599 and 1988, the 1951 eruption led to the creation of a formal authority, reorganized as the Philippine Institute of Volcanology and Seismology in 1984. Since experience had shown the importance of educating the public to the significance of early warning signs, a volunteer observers' programme was established in 1986 (De los Reyes, 1992). The Philippines also suffers from regular cyclones; between 1960 and 1981 there were 39 violent storms, which killed 5650 people. However, this figure, serious as it is, represents only a tenth of the number killed by cyclones in Bangladesh during the same period, mainly, it appears, as a result of the Philippines having more effective early warning systems (Wijkman and Timberlake, 1984, pp. 78–9).

SOIL

The wealth of a country, the crops it can grow and the animals it can feed all depend in the first instance on the nature of its soil. But soil is a very complex mixture, and the way in which it is used and the extent to which it is cared for are crucial.

Soil is created in the first instance by the erosion of surface rocks by wind and rain. Chemical weathering of minerals adds nutrients from the air; there is also loss from the leaching of chemicals from the soil. When vegetation takes hold, humus is deposited and plants, fungi and animals, particularly earthworms, ants and termites, contribute to the store of nutrients available (Trudgill, 1977). Grasses and trees not only bind soil together, but become the agent by which the valuable nutrients are recycled and slowly increased. Chelates released by the decay of organic matter act to help break down minerals. Rivers transport vast quantities of silt and spread it over large tracts of countryside. The process of soil formation is slow. At best, even when sediments build up quickly, formation of 30 centimetres may take 50 years. More usually, when new soil is formed from parent rocks, one centimetre may take from 100 to 1,000 years' (Myers, 1987, p. 24).

Soil, moreover, is not a constant commodity: it varies from intensely fertile to almost valueless in human terms, and this is where the political problems begin. The UN Food and Agriculture Organization (FAO) divides the uses of land into four main categories. The total land area of the world, 13,077 million hectares (131 million square kilometres), is classified as follows:

1. *Arable*: land suitable for growing crops, including permanent crops such as coffee and fruit, whether in use or not – 1473 million hectares (11%).
2. *Permanent pasture* – 3162 million hectares (24%).
3. *Forest and woodland* – 4090 million hectares (31%).
4. *Other*: waste land, including both urban land and desert – 4353 million hectares (33%).

Since 1945, the overall amount of arable land in use has increased by about 10%. During the same period, the population of the world has doubled, so that the increase in food production has been brought about by greater 'efficiency' in the use of existing land. In the same period a substantial amount of previously arable land has been lost to urbanization. A more worrying aspect is the unequal distribution of arable land between countries. In 1982, Brazil had 58 hectares of arable land per 100 inhabitants, but China and Bangladesh had 10 and Egypt only 6 (Cole, 1987, pp. 25–6).

Urbanization has been a major reason for the loss of arable land in Egypt; Cairo is now one of the largest cities in the world and one of the most densely populated. From an environmental point of view, however, the most important thing is that, by one means or another, the world is losing 7 million hectares of fertile land each year through soil degradation. Soil erosion is the most obvious form of degradation, although the degeneration of the soil through overuse is no less important for being more insidious. Some erosion is inevitable, especially in exposed areas. Water and wind erosion occur naturally, but the accelerated loss comes as the consequence of human activity and a variety of forms of mismanagement. Overgrazing, followed by deforestation, are the most important causes worldwide. Worst of all, 'soils can be degraded in a fraction of the time they take to form' (Myers, 1987, p. 24).

'The factors which influence the rate of erosion are rainfall, runoff, wind, slope, plant cover and the presence or absence of conservation measures' (Morgan, 1986, p. 1). It is vital to note that conservation does not mean stopping erosion altogether, but simply 'reducing the rate of soil loss to approximately that which would occur under natural conditions'. Worldwide, the main areas at high risk of erosion are:

1. *Semi-arid, semi-humid areas*, especially in China, India, the western USA, Central Asia and the Mediterranean.
2. *Mountainous terrain*, e.g. much of the Andes, Himalaya, Karakoram; parts of the Rockies, the African Rift Valley.
3. *Volcanic soils*, e.g. Java, New Zealand (South Island), Papua New Guinea, parts of Central America (*ibid.*, p. 2).

As can be seen from this, many of these areas are in the South, and much of the erosion is due to the intervention of human beings. Three factors determine how heavily land is used. The first is the perception by human beings of its value, whether for agricultural or other purposes. The second is the land's carrying capacity: the number of human beings it can sustain under normal cultivation. The third is the risk of crop failure, which leads traditional farmers to diversify their crops rather than to pursue high yields.

The low carrying capacity of the land in many Southern states has meant that, on the one hand there has been a great deal of land clearance, exposing the soil to the elements, and, on the other, the remaining soil is quickly exhausted by being overworked. In mountainous regions such as the Andes, the thin soil is barely adequate to sustain the human population and malnutrition is widespread. These lands are regarded as 'fragile' although fragility is not a quality that is easily measured and some fragile lands can, in fact, be highly productive (Browder, 1989, pp. 22–3). The result is that in Colombia, for example, 75% of all land is affected by erosion, so the fact that by 1982 there were already only 17 hectares of arable land per 100 Colombians gives grounds for concern. In Peru, where some 30% of the land is officially regarded as subject to erosion, the fast flowing rivers carry huge burdens of sediment as a witness to how rapidly the intensively cultivated soil is being washed away.

In the humid tropics, there are special problems. 'Where the land is cleared, erosion is highest in the rain forest areas whereas under natural conditions it is highest in the dense savanna areas' (*ibid.*, p. 3). The risk of erosion throughout Malaysia, for example, is at least moderate. However, much of the eastern side of peninsular Malaysia (Kelantan, Trengganu, Pahang, Johor) is at severe erosion risk. Much of the area is suitable for neither agriculture nor timber production. Blaikie explains why both continue to take place:

Inequalities between the majority of the rural populations affected by soil erosion and other more powerful groups are both a result and a cause of soil erosion. In this sense soil erosion is a symptom of underdevelopment and it reinforces that condition. Therefore to the rural poor at least, soil erosion is an important element in their poverty.

(Blaikie, 1985, pp. 3–4).

Of the world's 1.2 billion hectares with moderate, severe and extreme soil degradation, the largest areas are in Asia (453 million hectares) and Africa (321 million hectares). But Central America has the highest percentage of degraded land and the worst degree of degradation. Some 35% of the world's land is already degraded. As more marginal land is brought into cultivation, the rate of loss increases, typically exceeding the rate of soil formation by a factor of ten (Pimentel *et al.*, 1987). And there are other consequences of erosion. For example, the erosion of the highlands of India results in higher rates of sedimentation in the great rivers, which in turn makes the risk of flooding downstream and in the Ganges Delta markedly worse.

Blaikie (1985) points out, however, that identifying soil erosion as an environmental problem has since colonial times led governments to ignore the social problems that give rise to erosion. The users themselves are blamed for mismanaging their environment. Over-population is seen as contributing to their ignorance, as well as to the deterioration of the land. Lastly, they are seen as being insufficiently involved in the market economy, which would impose on them the discipline needed to use the land efficiently.

Against this, he suggests that small farmers mismanage the land not because they are ignorant, but because they are desperate. They farm marginal land, of poor quality or steeply sloping, because political decisions have left the best lands in the hands of large landowners or TNCs. For the same reason, they have little choice but to squeeze what they can out of what is left to them. Although they may manage their own estates well, therefore, agriculture and ranching companies contribute indirectly to soil erosion; logging companies and others, moreover, contribute to it directly, and on a much larger scale. Hence, the very market in which the technocrats seek salvation relentlessly drives the small farmer to degrade the land still further (Blaikie, 1985, pp. 138–40).

Degradation affects the least powerful people: small rural producers, pastoralists and, in towns, the urban working class, the old, the sick and the unemployed. To some it exemplifies the difficulty of balancing the demands of development and environment. However, it should not be forgotten that the basic mechanism of soil erosion is not only a problem for the South. Soil is also being lost in developed countries: e.g. in Norfolk, the UK, where the top layers are slowly blowing away into the North Sea. The only difference is that, up to now, Northern countries have been able to afford the high energy input needed to keep such soils fertile. The soil is still being lost to future generations. Poverty in the South accelerates this process, because soils break down more rapidly when overused and heavy tropical rains wash away the soil more quickly.

LAND OWNERSHIP

The most important thing about land is who owns it. This is not always certain and, as recently as 1954, no one at all owned about a third of all the land in the Central American nation of Honduras. However, in most countries the state is the residual landowner.

Unless all land is formally owned by the state, the pattern of land ownership will be unequal: most of the land will be owned by a few people and a large proportion of the population will have no access to land at all. Inequality of landownership is not, of course, confined to Southern states. Australia, an AIC, has one of the most unequal patterns of land distribution in the world. However, inequalities of land ownership tend to be particularly serious in Southern states for three reasons:

- They embody patterns that were imposed on the local inhabitants by conquest and colonization and are resented as such. This resentment is still being resolved: for example, in 1997, President Mugabe of Zimbabwe ordered the expropriation of the large estates still owned by European settlers or their descendants.
- Those who do not have land have to work for large landowners, or for other employers, in order to live. This can lead to problems when such employment ceases to available: in the absence of alternative employment, the tin miners of Bolivia, for example, have either gone hungry or have had to take to growing coca in order to stay alive.
- The particular structure of the global economy increases the pressures on the developing states to degrade their soils. For them, agricultural earnings traditionally form the basis of their export economies. And frequently they rest on few products, and these show a strong regional concentration.

Two-thirds of the exports from Africa south of the Sahara are coffee and cocoa. Nearly three-quarters of all exports from Latin America consist of three crops: sugar, coffee and soya beans. The situation is even worse for certain individual countries, whose reliance in extreme cases may be on a single crop: in the case of Cuba, sugar, and of Bangladesh, jute. Indebtedness further increases pressures to export, and what is produced will then be determined by the global market place.

The move into cash crops takes up much of the best, most fertile land, and subjects it to intensive cultivation of a single crop. Traditional patterns of crop rotation, which have kept the land in good condition for decades, if not for centuries, are abandoned, fertilizers and pesticides are imported and used to compensate for lost fertility and this intensive use can lead to soil degeneration. Furthermore, there are significant political consequences. The move into cash crops is generally accompanied by a concentration of land ownership. Even where traditional peasant cultivators are not forcibly deprived of their traditional lands, or find them taken from them by the abuse of the legal process, they often find it hard to compete in the global market place. 'Urban-biased policies seldom give much support to peasant cash-croppers: market prices are allowed to fluctuate or tend to be kept so low (to satisfy the city-dweller voters) that the producers get inadequate, insecure returns and may over-exploit the land to survive' (Barrow, 1995, p. 115).

The Central and Eastern Forest Region of Côte d'Ivoire, the 'Cocoa Belt', has suffered serious soil degradation, and this has been exacerbated by declining terms of trade for local producers of coffee and cocoa. Again, the interaction of the global market system and state economic structures exacerbates economic problems. As elsewhere, falling returns on cash crops tend to lead to more extensive production and thus to further forest clearance, as well as more intensive use of existing agricultural land. On the other hand, cocoa production is much less damaging to the environment than, say, cattle ranching, since the trees rapidly form a canopy and so continue to afford protection to the soil below. Cocoa is in fact one of a number of tree crops, such as brazil nuts, sôrva latex and piassaba fibre, that occur naturally in the Amazon rainforest, and that could be systematically developed in a sustainable fashion (Prance, 1989, p. 67). However, this can only happen if a systematic effort is made to find markets for them.

Where agricultural use necessitates irrigation – and 12% of the world's cultivable land is irrigated – this increases the degradation of the soil. Some 50% of Thailand's land resources are subject to erosion, as a result of deforestation, or acidity and salinity, as a consequence of the overuse of irrigation water. In Pakistan, more than 65% of the land is affected by salination/waterlogging. In the Indus Valley, cultivated for the past 5000 years, crop yields are now among the lowest in the world. In Egypt, 35% of the land is irrigated. Large landowners can use their political power to gain access to irrigation water and to deny it to others who are less fortunate. This process in Egypt, and the very high levels of unemployment there, have helped to fuel 'Islamic' militancy.

The reasons for soil degradation are the same the world over, but their proportionate contribution varies a good deal by region. But, whatever the reasons, degradation may lead on to loss of trees/ grassland and sometimes eventually desertification, a process which is reversible in its early stages. This process is illustrated by the American dustbowl, but is more obvious in the South: for example, in Iraq, Ethiopia and the Sahel region.

DRY LANDS AND DESERTIFICATION

At the United Nations Conference on Environment and Development (UNCED), held in Rio de Janeiro in 1992, desertification was not seen as a global problem in the way that climate change or biodiversity were (and are). The Convention on Desertification was a response to African pressure and was only signed subsequently. The poorer African states of the Sahel had been calling for global action on desertification throughout the 1970s and 1980s. There was no doubt about its importance to Africa: television had beamed pictures of starving children in Ethiopia, Sudan and Somalia on to television screens throughout the developed world. What was at issue was its urgency as a global question.

The OECD formed the Club du Sahel in 1976 in order to combat desertification and increase food production in the dry zone. Responding to the crisis in the Sahel, the UN General Assembly called a Conference on Desertification (UNCOD), which met in Nairobi and agreed on an international Plan of Action to Combat Desertification (PACD). However, the PACD immediately ran into problems that, as is shown in Chapter 8, are common to all attempts to organize international responses to environmental problems. The plan stated what should be done but not how it was to be achieved, and from the beginning it was underfunded. Part of the reason for this was an organizational problem: the Plan had been generated by and remained under the supervision of the UN Environmental Programme rather than the UN Development Programme or FAO, both of which command much more political support and with it resources from the AICs (Barrow, 1995, p. 118). Since the 1980s, moreover, debt problems have again diverted funds in individual countries away from lands which are seen as 'marginal' towards increasing production in the more fertile lands. Desertification is certainly an international problem, and not just because it affects both climate change and biodiversity. Migration due to desertification may, for example, become a destabilizing factor, with repercussions for the North. Areas subject to desertification include:

- Kuwait, which is arid rather than semi-arid. In the Abdaly region, 156 farms are separated by windbreaks and desert, and are subject to encroachment.
- The central area of Argentina, affecting some 16 million hectares. In South America generally, some 22% of arid regions are severely or very severely affected.
- China, where some 30% of approximately 17 million hectares of desertified land has been created since 1920. Sand now extends some 70 kilometres south of the Great Wall, although protective forest systems created since 1978 have had some success in halting the advance and creating oases (Mannion, 1991, p. 228).

The UN Plan of Action to Combat Desertification (PACD), instituted by UNEP and approved by the General Assembly in 1977, was never effectively implemented, largely because of the lack of available resources. There was some sense of trying to confront issues beyond human control. Despite radical suggestions about the human causes of desertification, the characteristics of areas such as the Sahara predate human life (Goudie, 1986, p. 51). But reluctance to support PACD was reinforced in 1992 by new satellite pictures that showed that the general advance of deserts, then widely accepted, was a myth (Pearce, 1992). This is an important example of the way in which hostile interests can make use of scientific evidence to halt or postpone action. The Sahel is a transitional region, subject to irregular variations in climate, so there was conflicting evidence from different regional studies, which could be used as an excuse to postpone the decision, as the North wanted. But 'wait and see' policies seemed to be in fundamental conflict with the prevailing ideology of Rio, the 'precautionary principle', which was emphasized where it suited the developed nations.

Land degradation due to lack of water may possibly be caused by climate change, but drought alone is not the real problem, despite the breaking down of established systems for coping with drought in many parts of the South. Human action can take dryland degradation right through to desertification as a result of a combination of human activities with an occasional series of dry years. Although evidence from North Africa following the Second World War indicates that the desert has the capacity to recover, it does so exceedingly slowly and only if not further disturbed.

The conversion of land to grazing is also regarded as a cause of desertification. Grazing has a range of effects on plant species, and not all of them are bad: some species benefit from grazing. However, stock numbers have risen in many dryland regions. Between 1955 and 1976, cattle numbers in LDCs rose by 34%, and the number of goats by 32% (Arnon, 1981, p. 111). Moreover, stable patterns are not necessarily best for drylands: migration traditionally reduced losses in bad seasons, and the new patterns of land ownership restrict these (Barrow, 1995, pp. 113–14).

The Kenyan drylands are an example of these desertifying processes at work. There, the ability of the land to sustain its sparse and nomadic population is increasingly threatened. Overgrazing is occurring because there is less and less land available to herders. Land has been sold and fenced for settled farming, especially for the growing of cash crops. However, the effect of this has not necessarily been harmful. Such improbable crops as roses generate necessary foreign exchange in indebted and 'structurally adjusted' countries where terms of trade favour the North.

Some environmental problems need top-down global solutions and top-down efforts need coordinating. There have to be concerted policies by individual countries and international NGOs. But a bottom-up approach is also necessary, perhaps more so for some problems, and desertification is an example. Local people must be involved, local conditions must be addressed. Empowerment is vital to deal with desertification; since the most obvious signs of the problem are local, local solutions are needed. Small-scale soil and water conservation activities have led to considerable environmental recovery in parts of Kenya.

DUMPING OF TOXIC WASTES

The October 1997 edition of *New Internationalist* was concerned with trash. Vanessa Baird, in the Keynote article, writes: 'The sociology of trash is simple: the rich make it, the poor deal with it.' This is, of course, simplistic, but it contains some truth at various levels – local, national and international – and to various degrees of threat, from litter to toxic waste dumping.

An increasing problem for the South is the disposal of toxic waste. Industrial processes, by their very nature, are likely to generate toxic by-products. The problem is, first, that up to now it has been assumed that the benefits of these processes outweigh the disadvantages, and, second, that the obvious way to get rid of waste is simply to dump it in large piles or plough it into the ground.

The irony is that most Southern states have been generating relatively little toxic waste. The major sources are the AICs and the NICs. 'Industrial economies typically produce about 5,000 tons for every billion dollars of GDP, while for many developing countries the total amount may only be a few hundred tons. Singapore and Hong Kong combined generate more toxic heavy metals ... than all of Sub-Saharan Africa (excluding South Africa)' (World Bank, 1992, p. 54; cited in Middleton et al., 1993, p. 152). Nor do the citizens of the South in general generate as much 'municipal' waste: that is, the everyday waste produced by individual households. Not only are they much more likely to 'make do' and make the best possible use of everything that comes into the home, including packing materials, but if they do throw it out, there are large numbers of people who make a living by scavenging on the municipal tips. The process may look ugly and it is certainly hazardous for those who do it, but scavenging does work, recycling materials that would otherwise become landfill and reducing the overall impact of consumption on the environment.

However, future development is likely to change all that. In 1969, Thailand had only 500 factories producing hazardous wastes; by 1993, it had 26,000 (Middleton et al., 1993, p. 154). The story has been repeated not only in the NICs but in most other Southern states. Lax controls, resulting from both corruption and the overriding imperative to compete in the world market by limiting demands on industry that would increase production costs, mean that no reliable estimate of the amount of harmful waste generated worldwide exists. The figure of 375 million tonnes per year, reported to the Brundtland Commission, is likely to have been much too low (World Commission on Environment and Development, 1987, pp. 226–7).

The preferred option in the AICs has been a combination of recycling and incineration. Recycling means requiring industry to reprocess waste products to reduce their volume and toxicity as far as possible, and calling on the householder to separate out different types of waste. In the South, the former is little tried as yet, although, as noted above, the latter does tend to happen anyway.

Incineration that does not involve the use of the heat generated to produce electricity is itself wasteful, and by releasing carbon dioxide more quickly, impacts on the environment through the greenhouse effect. Substantial investment is therefore required to make the process efficient, and the incinerators have to be situated close to residential areas in order to make the best use of both the electricity and the waste heat that is inevitably generated. However, this is not easy to do, since incineration is the preferred option, especially for the destruction of complex organic substances. These require very great heat to ensure that the combustion process is complete and that no toxic waste products remain. Concern that this cannot be guaranteed has led to public protests at the siting of incinerators near residential areas, and to date no satisfactory resolution of the dilemma has been found. Certainly, such advanced facilities are unlikely to be available in most of the South for a long time to come; in the meantime much waste will continue to be burnt on open fires, without any precautions at all to avoid contamination.

MINING

Deep mining has become increasingly uneconomic. The mining of tin in Bolivia, for example, ceased altogether with the collapse of the International Tin Agreement (Crabtree et al., 1987). The modern approach is even more destructive to the environment — this is strip mining or open-cast mining, where the entire top surface is removed. This method produces half the coal output of the USA, and enables Colombia and Australia to beat the cost of UK coal at UK ports, exemplifying the neoliberal tendency to ignore externalities such as environmental damage in assessing productive costs. Strip mining is also increasingly the method used by the international mining companies, which are often exploring and mining in the world's most fragile environments: the Pacific Islands, for example.

In Papua New Guinea, copper mining by RTZ/CRA, a British/Australian company, has produced more than a billion tons of waste and killed all aquatic life in a 480 square kilometre river system. UK, Australian and Canadian gold mining companies, including CRA again, while exploiting the alluvial gold in the southern uplands of Papua, are discharging large quantities of cyanide-contaminated waste into an area of ocean renowned for marine biodiversity. Australian mining of nickel (named after Old Nick and a class 1 carcinogen) and cobalt is also a threat to coastal environments in the Pacific. US oil companies are also just beginning to exploit Papua New Guinea. It is clear that the lessons of past destruction have not really been taken on board by these companies, despite dead rivers and massive environmental and human costs of other kinds. Kennecott, a division of Rio Tinto, is beginning mining for gold in one of the smallest, and hitherto unspoilt, Papuan islands. The case also powerfully illustrates the vulnerability to environmental damage of the lifestyles of indigenous peoples.

Phosphate mining has reduced two-thirds of the island state of Nauru to wasteland. Other islands, notably Banaba, have been totally destroyed or had their lagoons poisoned. Gold mining in the Solomon Islands is damaging a sensitive ecology and causing displacements of people. The future for the Pacific states does not look any better. A proposed copper mine in Fiji, exploiting marginal low-grade deposit, would dump 98,000 tonnes of poisonous waste into the sea every day. The Canadian and Australian TNCs involved are supported by the Fijian military and members of the Council of Chiefs, who have controlled Fijian politics since the 1987 coup. More than 20 companies have licences for further exploration.

In Amazonia, gold miners (*garimpeiros*) are called 'earth eaters' by the Indians. Their violence is legendary. That, along with their ecological impact, has produced a backlash from the Indians, which has been met by massacres (Albert, 1994, pp. 49–50). As a result of its use to separate gold from alluvium, many of the tributaries of the Amazon are polluted with mercury, which poisons the wildlife and the people who feed on that wildlife. Laws exist to control the sale and use of mercury in Brazil, but they do not have much effect in a frontier situation. The good news that the number of *garimpeiros* is decreasing masks the increasing scale of the gold mining in the region, but this in turn still pales into insignificance against the effects of some official mining projects in the area.

Since its commencement in 1980, the Projeto Grande Carajás (PGC) in Brazil has become a massive multifocal programme centred on the Serra dos Carajás, a mountain range in the centre of Pará State (Hall, 1989). In 1967, as a result of the chance landing of a helicopter owned by the US Steel Corporation, the Serra dos Carajás was discovered to contain one of the world's largest deposits of high grade iron ore (at least 20 billion tonnes), as well as huge deposits of copper, gold, bauxite, nickel, manganese, potassium and other metals. The discovery led to the putting together of the PGC, which began in 1980 as a partnership between US Steel and Companhía Vale do Rio Doce (CVRD), a Brazilian company funded by both private and public capital. However, US Steel later pulled out of the collaborative project and subsequently the Japanese foreign aid body, JICA, became the main partner and outside financier. Today CVRD is the biggest exporter of iron ore and the third largest mining company in the world, while the PGC has become an immense industrial complex with a large number of spin-off operations. It includes a 900 kilometre long railway, which carries the yield of the Carajás mines across the Amazon jungle to São Luís in Maranhão State. At São Luís there are a large number of Japanese and US-owned aluminium smelting plants, as well as the deepwater port facility of Porto de Ponta da Madeira, from where large amounts of iron ore are exported to Europe and the United States. In January 1996, surveyors for the CVRD discovered a huge deposit of gold in the Serra Leste, which has further enlarged the area of the Amazon occupied by the workings of the PGC (*Newsweek*, 19 February 1996, p. 39).

Mining of all kinds involves the moving of vast quantities of material. Overburden and waste are not always replaced, although sites may in some instances be used for 'landfill' (dumping waste) before being covered. The impact on the environment of Colombian coal mines or Peruvian copper mines, however, is dwarfed by the total impact of the exploitation of sand and gravel, crushed limestone (lime), aggregates, etc., which are the essential raw material for modern concrete construction and road building. The annual movement of soil and rock from all causes worldwide is estimated to be as high as 3000 billion tonnes (Holdgate *et al.*, 1982, p. 186). The results include subsidence, the interruption of watercourses and the accelerated sedimentation of rivers and oceans. By comparison, best estimates of all the sediment carried into the ocean by the world's rivers each year is just 24 billion tonnes.

THE ATTACK ON THE FORESTS

Since 1850 the world's forests have declined from 6 billion hectares to 4 billion, and the rate of loss is accelerating, even in the years since the Rio Summit. At the beginning of the twentieth century, more than half of India was covered in forest; now the figure is 14% (Glasbergen and Blowers, 1995, p. 5). Forests are threatened by four main causes.

- *Logging*. Logging leads to loss of forest cover in two ways: directly, the loss of the trees themselves; and indirectly, by the practice of 'clear felling' (removing all that stands in the way of the desired species) and the opening up of wider tracts of forest to settlement by the creation of clear spaces and access roads. Many of the forests of the Philippines, Malaysia and Indonesia are effectively logged out.

- *Cutting for fuelwood.* Many Southern cities have 'fuelshed' regions extending up to 300 kilometres from their urban limits. In Sierra Leone, as the zone of exploitation spread further and further around Freetown, attention turned to the mangrove trees that protect the coastline.
- *Land clearance.* Land colonization or resettlement is a policy regularly employed by governments to avoid pressure for land reform. Vacant land resettlement programmes in Indonesia, Amazonian Brazil, Peru and Ecuador, Bolivia, Malaysia and various African countries 'have generally caused deforestation and land degradation' (Barrow, 1995, p. 116).
- *Clearance for military security.* The extreme case was the regular deforestation of Vietnam, Cambodia and parts of Laos during the Vietnam War by US forces using 'Agent Orange'. The Mexican government, however, is currently encouraging clearance of forest and settlement in Chiapas to forestall its use by guerrillas.

Most forest loss in the past has been in temperate zones, but in recent years many Northern countries have taken steps to conserve and even to extend their renewable resources. The tropical rainforest was largely spared until quite recently because of distance and the difficulty of access. But the acceleration of forest loss is now fastest in the tropical regions, where 60% of the remaining forests lie and international protection is minimal. The International Tropical Timber organization established in 1985 and sponsored by the UN never got off the ground, as importing nations refused to pay the fees that should have gone to it, and Japan boycotted its meetings.

The idea of the forests as 'global commons', sustaining us all through the provision of oxygen, the processing of pollutants and their role as a home for diverse biology, is not necessarily the way everyone sees it. For example, Mahathir Mohamad, Prime Minister of Malaysia, in 1992 accused environmental groups trying to constrain the Malaysian timber trade of 'advocating that forest dwellers remain in the forest, eating monkeys and suffering from all kinds of diseases' (Tobin, 1994, p. 290). His view, put bluntly, was that if the developed world wanted Malaysian forests it had to be prepared to pay. It is understandable that an NIC like Malaysia should view forests as a resource to be exploited, but despite the Mahathir suggestion that the developing countries are entitled to exploit their 'own' resources, it is actually the developed countries that import and thus consume these resources, and international logging companies that have contributed most to the gathering disaster.

In Africa, 40 years ago, 30% of Ethiopia was covered by forest; now forest covers less than 1% of its territory. Deforestation was a major cause of the droughts of the 1970s and 1980s. Côte d'Ivoire has lost 75% of its forest cover since 1940, and the rate of loss is estimated at 6.5% per annum, one of the highest in the world. Public expenditure cuts have led to cuts in the agencies managing the forests and in replanting projects. National rates of loss may in some cases obscure massive losses of some of the most important forest areas. In Kenya, the Imenti Forest around Mount Kenya was burned in 1997 by illegal settlers while forest officials turned a blind eye. In the case of Rwanda, where huge tracts of the Nyungwe Montane Forest Reserve were left burning for weeks, threatening the very survival of the cloud forest, government weakness and local incapacity were also part of the explanation.

In Asia, Thailand, declared a disaster area by the Worldwide Fund for Nature (WWF) in 1989, has less than 20% of its primary forest left. But banning logging in one place following popular disaffection does not solve the problem, as even when this has been done, the companies have simply switched their operations to somewhere else. In the case of Thailand, they have moved either to Burma, aided covertly by military interests there, or to Laos, where the central government is so weak that it is unable (even if willing) to prevent the destruction.

In the Philippines during the 1960s, President Marcos encouraged the exploitation of the forests and 80% of the timber logged was exported. It was soon apparent that the logging was unsustainable, and in 1975 a quota of 25% exports was introduced to encourage domestic processing of timber; every male over ten years of age was required to plant a tree a month for five years. It was too little and too late. By 1976, high grade logs were exhausted. By the 1980s, the problem of a shortage of fuelwood was exacerbating forest damage. In 1986, Corazón Aquino recognized deforestation as one of the Philippines' main problems and began to take serious action. National funds as well as aid from the Asian Development Bank and Japan flowed into social forestry projects and replanting schemes, which were also supported by NGOs, including church groups. By the late 1980s, harsh laws against illegal logging were in place and some land reform had been introduced.

The islands of the South Pacific have become an alternative source of supply for the markets of Japan and Korea. The Japanese have been very careful to conserve their own forests, seeking instead alternative supplies in South-East Asia. Much of this wood is turned into chipboard or used in the manufacture of pallets on which to stack Japanese exports of electronic goods. They are used once and then burnt, further contributing to global warming.

Malaysian logging companies are in turn ruthlessly exploiting the forests of Melanesia. Village communities have lost their traditional lands in Papua New Guinea and the Solomon Islands as the logging companies have 'bought' those lands from officials purporting to have the right to sell. Logging in the Solomon Islands is proceeding at such a rate that the islands are expected to be logged out in five years. In Papua New Guinea, speculators have taken advantage of the worst drought for 50 years to burn thousands of square miles of forest and grassland, leaving millions of its inhabitants on the edge of starvation. The loss of cover is so great that it is expected to affect the climate adversely, making matters still worse.

Logging in Fiji was welcomed by its rulers, since Fiji's refusal to accord equal treatment to its large minority of Indian origin had made it a bad investment risk. Now Fiji is virtually bankrupt despite the exploitation of its resources, and sought readmission to the Commonwealth in the hope that it may once again qualify for support.

A contributory factor, especially in Indonesia, has been the clearance of swamp forest under government-sponsored schemes to grow rice, nominally to feed Indonesia's 200,000,000 human inhabitants, although it seems very unlikely that the land is ever going to be very suitable for that purpose. In 1997, an area the size of the UK burned for weeks in Sumatra and south and central Kalimantan, creating a vast cloud of smoke (euphemistically referred to as the 'haze'), which blanketed Indonesia, Singapore, Malaysia, Southern Thailand and part of the Philippines (*Keesing's* 41823–4, 41869). Five million acres of forest were lost. Although the later arrival of the monsoon checked its further spread, the fires continued to burn underground in the peat subsoil, and reappeared early in 1998.

In December 1997, the WWF released a report which fully confirmed the seriousness of the situation. The year 1997, it said, had been the worst in recorded history for the destruction of the environment, and the main agent had been fire. As always, some forest fires had natural causes, but the damage done by lightning has had much more serious consequences as a result of selective logging, drainage and possibly also climate change. Accidental fires, too, resulted from causes such as charcoal burning, cooking and the dropping of lighted cigarettes. Claude Martin, director-general of the

WWF, was unequivocal in saying that the main damage had been done by fires 'deliberately set' by human beings, and the main reason was the general breakdown of law and order, leaving people free 'to clear forest for other purposes, and sometimes to cover up for timber poaching and land theft'. He added: 'It is also increasingly clear that the guilty parties are not just small farmers, as has been assumed, but most of the damage is being done by plantation and timber companies. This is not just an emergency, it is a planetary disaster' (Brown, 1997).

As noted above, national rates of loss may seriously underestimate the loss in specific regions. For example, the FAO gives a national rate of 1.3% for Mexico, but regional reports suggest that losses may be as high as 25% in some areas (Reed, 1992, p.194). Mexican government policies for the alleviation of poverty through allocation of plots of land, supported by well intended World Bank sectoral loans, have contributed to the destruction of Mexico's forest. Some 20 million hectares of tropical rainforest are lost each year.

In Colombia, there were 7000 forest fires in 1997 alone. Of these, 37 occurred in areas which were supposed to be protected, resulting in the damage of some 16,200 hectares of national parks.

There has been particular interest in deforestation of the Amazon Basin, as the world's largest land-based carbon sink. For successive Brazilian governments, however, the Amazon Basin has been seen as underdeveloped territory that they can use to alleviate the pressures of a growing population without having to carry out land reform. Even more crucially, settlement in Amazonia is seen as a necessary prerequisite to developing its resources for the good of Brazil and the fulfilment of its future destiny as a great power.

The framework for present Brazilian government policies on the Amazon region was decisively shaped by the 21 years of direct military rule that Brazil experienced between 1964 and 1985. Following their seizure of power in the so-called 'Revolution of 1964', the military's leaders were determined to turn their country into a great power (the catchphrase 'Brasil Grande Potência' was a popular slogan of the military government), commensurate with the country's great size and abundance of natural resources. In order to achieve their goal, the military put together a strategy of national development, known as the Plano Nacional de Desenvolvimento, which owed much to geopolitical considerations (Hepple, 1986).

The Polonoroeste ('northwest pole') and Polonordeste ('northeast pole') projects were conceived as two complementary 'development poles', which would, in the words of President-General Médici, 'give to the land-less people of the Northeast, the people-less land of the

Amazon'. The two colonization projects were directed by INCRA (Instituto Nacional de Colonização e Reforma Agrária), which, despite its title, aimed to defuse social tensions in the North-East without resorting to agrarian reform. In the 1970s, INCRA carried out intensive advertising campaigns with the aim of encouraging the North-East's impoverished peasants to migrate to plots of land situated along the newly built Amazonian roadways. At the same time, cattle ranchers were attracted to the region by the stimulus of official incentives and subsidies. The Projecto Polonoroeste was particularly destructive of the region's ecology, and it was largely as a result of the project that a quarter of the original Amazon rainforest had been destroyed by the early 1970s (Guimarães, 1991, p. 132). The project principally affected the state of Rondônia and was centred on the BR-364 road, which cut across a large section of rainforest on its way from Cuiabá in the state of Mato Grosso to Rio Branco in the state of Acre. The consequent rush of migrants to the region resulted in the deforestation of extensive parts of the state of Rondônia, and one-sixth of the state lost its natural cover as tracts of land were cleared by settlers and ranchers, either to make way for cattle pastures or for cultivation purposes (Guimarães, 1991, p. 221). The deforestation process was greatly accelerated by the decision of the World Bank to fund the asphalting of BR-364.

The longstanding resistance of internal elites (especially those with interests in the Amazonian mining and ranching industries) to what they term the 'internationalization' of the Amazon – that is, the involvement of the outside world in what they believe is a matter of national sovereignty for Brazil – has been one of the main reasons why the Brazilian government has not in recent times taken firmer action to protect what remains of the Amazonian rainforest. International concern over the destruction of the rainforest grew during the late 1980s, when aerial surveys revealed that about 372,700,000 square kilometres or 9.3% of the original Amazonian rainforest had been deforested, and it was realized for the first time that the burning of the forests was making a substantial contribution to the global 'greenhouse effect'. As a result of protests by international environmental organizations in alliance with Brazilian ecologists, Indians and rubber-tappers (whose leader, Francisco 'Chico' Alves Mendes, became the world's most celebrated 'eco-martyr' when he was assassinated in December 1988), the Brazilian government announced in October 1988 that it would restrict subsidies for agricultural developments in the Amazon region, prohibit wood exports and strengthen the relevant environmental agencies. However, the large number of qualifications added to these measures meant that the government's actions were given a rather cool reception by the opponents of deforestation.

The United Nations Conference on Environment and Development was particularly important for focusing much of the world's attention on the ecological consequences of deforestation in the Amazon Basin. At UNCED, the Brazilian President, Fernando Collor de Mello, took the opportunity to tell the world just how much Brazil (and by implication its President) was doing about the environment (Collor de Mello, 1992). However, the 'green' credentials of the President were in doubt, especially after he dismissed the renowned environmental activist José Lutzemberger from the post of environment minister, when the latter had accused the governmental agencies, which had been ostensibly established to police illegal clearing and logging activities in the Amazon, of being 'dens of thieves, wholly-owned subsidiaries of the international timber companies' (*Folha do São Paulo*, 22 March 1992). Collor de Mello was subsequently successfully impeached and, despite his government's backing at UNCED for wide-sweeping treaties to protect the environment, the Amazonian rainforest has continued to suffer deforestation, with 1995 having witnessed a renewed surge in forest burnings following an improvement in the Brazilian economy's growth rate (*Newsweek*, 8 January 1996).

Brazil is currently the world's fifth largest timber exporter, and is likely to become even bigger in the field as Asia's tropical forests are depleted. Brazil is a major exporter of mahogany and the largest supplier to the UK. It resisted the inclusion of mahogany as an endangered species whose international trade must be monitored at the meeting of the Convention on International Trade in Endangered Species (CITES) in June 1997, which also saw the leak of a document from the Brazilian Secretariat for Strategic Affairs, which admitted that 80% of the timber extraction from the Amazon region was illegal and thus out of the control of the Brazilian government. Such extraction produces minimal rewards for the local community, is inefficient and totally unconcerned with sustainability. Despite the illegality of such operations, the logging companies, including 22 foreign-owned, mainly Asian, ones, continue to get the tax breaks which were established by the military government to encourage 'development'.

Brazil's sensitivity to foreign concern about the Amazon was backed by a meeting of the Amazon Pact countries on 6–8 March 1988, and was reaffirmed by President Sarney when, on 8 March, during a visit to Guyana, he stated that the Brazilian Amazon was a sovereign matter and complained that the criticism was turning the region into 'a green

Persian Gulf'. These views were echoed at UNCED in 1992, when Malaysia refused to sign the convention on biodiversity, asserting its right to 'exploit' (i.e. destroy) its remaining tropical rainforest if it chose, precisely as President Sarney had earlier done in Brazil.

Hence, there was no agreement on a forestry convention which would check the destruction in time to forestall irrevocable loss on a massive scale (it was forecast that Malaysia would have lost all the rainforest in Sabah and Sarawak in the following eight years). Instead, there was a vague Statement on Forest Principles and a call for governments to meet again to iron out the remaining difficulties.

PROTECTION OF FORESTS

Forests constitute 85% of the Earth's biomass, and their protection is vital. There have been some notable forest environment successes. The Chipko 'hug a tree' movement is perhaps the most famous environmental action of all, with its slogan 'the forest is our mother's home'. Women in Uttarakhand, India, left behind in the hills while their menfolk earned their livelihoods on the plains, were having to walk further and further to collect fodder and fuel due to shortages caused by commercial forestry. After an eight-year struggle, the Indian government imposed a ban on the felling of green trees for commercial purposes on hill slopes of 30 degrees or more and slopes above 1000 metres in the Uttarakhand Himalayan region.

The movement is often seen romantically, as a confrontation between two different ways of seeing forests: the spiritual and the utilitarian. But the utility of the forests to the women of the Uttarakhand region is in fact central to the dispute. In fact, the dispute exemplifies the different uses of forests in North and South. The developing countries traditionally used forests primarily for fuelwood, whereas the developed world uses timber primarily for industrial purposes (Agarwal, 1986). Alternatives to these patterns of consumption must be sought. For the most part, protection of forests has clearly had to rest on the policies of national governments, though they have frequently been supported in these by environmental NGOs. This protection must also fit together with the lifestyles and aspirations of local people.

In the Korup National Park in Cameroon, a rainforest preservation project, developed in the mid-1980s by the Cameroon government in conjunction with the WWF and 34 other institutions and NGOs, seeks to combine preservation with help to the local people. They are encouraged to meet their own needs from the forest without exporting forest products to Nigeria. Another approach is to seek to develop productive alternatives to timber. Some countries are looking at forest products suitable for export other than timber. Such alternative industry is encouraged by NGOs, which have been seeking to develop sustainable exploitation controlled by local communities and benefiting them.

But the protection of the forest is always going to be on a small scale if there are interests seeking to exploit it. One of the factors promoting such exploitation is indebtedness. This can be mitigated through debt for nature swaps. These involve debt reduction (usually dollar-denominated) in return for local expenditure (in local currency) on the local environment, but the problem remains of striking a fair price and avoiding further charges of exploitation and 'eco-imperialism'. Apart from cancellation of such debts, there are other actions that can be taken at both the local and international levels. In 1992, the Earth Summit agreed a Tropical Forest Preservation Programme, and US$1.5 billion was to be made available for it. The release of the second tranche was discussed at the G7 summit in June 1997, but the final decision was left to the donor countries to take at a special meeting in Brazil in October 1997 – five years after the original decision.

The problems consequent upon loss of forest cover do not just affect local environments through erosion, silting of rivers and the flooding of arable land. Forest loss degrades the global environment through loss of biodiversity and disruption of wider ecosystems by an increased contribution to carbon dioxide emissions and a loss of carbon sinks. It can be argued that deforestation is the result of relative poverty, while global warming is the product of affluence, but to do so is to ignore the intimate connections between all aspects of environmental degradation. It is also to ignore the massive loss of temperate rainforest, a problem much less widely discussed. The temperate rainforests constitute a living biomass twice as great as that of the tropical rainforests. Trees grow taller and live longer. Despite very effective local NGO activity, supported by networks of national and international groups such as the European Rainforest Movement, to protect the rich biodiversity of the temperate rainforests, exploitation continues on, for example, Canada's west coast. The government of British Columbia sees itself as successfully straddling a variety of interests, including the need for work of Canadian lumberjacks. It has produced regulations for how the forest is used, which have been tightened as a result of NGO pressure, but the valleys are still being stripped ('clear felled') and ecosystems disrupted. A popular backlash has led to belated changes in government forest policies, which now include charging logging companies for the damage they do.

CONCLUSION

Land comprises both soil and subsoil resources. In the case of the soil, the end of colonial rule has not brought any lessening of the pressure to exploit the land; on the contrary, indigenous elites are doing so with a ruthlessness that colonial powers often failed to show. The dominant problem of the South today is that economic pressures to use inappropriate technology involve the conversion of small-scale peasant cultivation to large-scale mechanized farming. Since the peasants themselves lack the capital to 'modernize' in this way, a variety of devices are employed to deprive them of their land and to concentrate ownership in the hands of a few individuals or corporations. This in turn makes it much more difficult for small farmers to stay in business. They need an additional source of income in order to survive, and by selling their labour to the large landowners they seek to recover a degree of economic independence, thus making available to the landowners the generous supply of cheap labour that enables them to drive down prices even further.

Driving peasants from more productive to more marginal land (or off the land) and into poverty presents fresh problems for the environment. Poor living conditions result in increased organic pollution of water and land, and the failure to eradicate malaria and the revival of tuberculosis in the South both clearly demonstrate the link between poverty and disease. Soil degradation stems from a combination of causes. However, in the South the common factor accelerating the degradation of the soil is poverty. There are serious doubts whether in any true sense the intensive agriculture of the North is sustainable. But in the South the ability to use high-energy inputs of fertilizer and pesticides is confined to the big landowners (including foreign-owned corporations). Elsewhere, the pressure is to try to squeeze as much as possible out of the soil with the minimum of input. The South gets the worst of both worlds. Land is degraded both by the excessive use of fertilizers and by their inadequacy. The desire to clear land for farm crops and ranching results in the destruction of forest cover and faster soil erosion.

Conversion of the land to agribusiness has wider consequences. Deforestation has now reached the point where it is belatedly being acknowledged as a global problem, but the remedies are still too slow and too weak. Meanwhile, the pressure to get a living from the newly cleared land leads to the overuse of fertilizers and pesticides, and in a few brief years to its degradation. However, the process cannot be arrested because of the power of the large TNCs and the main landowners, backed in many cases by foreign money, and the close relationship between their power in the countryside and that of the politicians in the national and provincial capitals.

With the end of colonial rule, too, the temptation to exploit the non-renewable resources of the subsoil has not lost any of its attractions. Mining appeals to the 'bonanza mentality' and offers the glittering prospect of a large pay-back for the local community in immediate returns. However, as with other primary industry (notably quarrying), mining also means spoil tips, pollution of the environment, etc. Usually being remote from centres of power, mines are little restrained by environmental legislation, even where it exists.

CHAPTER 6

THE BIOSPHERE

THE BIOSPHERE AND THE FOOD CHAIN

The biosphere is the whole interrelated biological system of which we are a part. It includes the land, its soil cover and forests, its diverse lifeforms, the whole range of animal and plant species and the waters of the earth, the rivers and oceans (Bradbury, 1991). Species have evolved over a very long time scale, much longer than human history. Once extinct, they cannot be replaced, although other species may move into the ecological niche vacated.

Although we are used to thinking of the natural world as divided into two kingdoms – the plant and the animal – it is more usual today for biologists to use the nature of the material from which cell walls are constructed to distinguish between four: plants, animals, fungi and insects. All four play an essential role in making human life possible; hence, on the precautionary principle, the only sensible strategy for our long-term survival (let alone that of the biosphere as a whole) is the conservation of all species.

A minority of paleontologists believe that human beings may at one time have been aquatic. Certainly since time immemorial human beings have found food on and near the sea shore. The concept of the food chain was originally developed in the context of the sea; it was only later realized that it applies on land and to all living creatures. It underlies all consequences of the use and pollution of marine resources. Fish depend ultimately on the pattern of tides and currents that bring concentrations of certain microorganisms to one part of the sea rather than another. The last major untapped resource, in the South Atlantic, is already under threat. As will be clear from the decline of specialist feeders such as the giant panda (see below), being located at the top of a short food chain is no guarantee of survival. Indeed, it may reflect a very vulnerable position. Food security for any species (including our own) rests on a complex and diverse food chain offering at each stage a wide range of choice.

HUMAN IMPACT ON THE BIOSPHERE

In the past, some species were deliberately hunted to extinction, at least locally. In heavily populated European countries, wolves and bears were exterminated because they were a threat. Today tigers are still deliberately killed in India for the same reason, although most such hunting is the result of the misguided belief that consumption of tiger genitals enhances human sexual prowess. Other species have succumbed to hunting. Not only is the Arabian oryx extinct in the wild, but the vogue for hunting in the Middle East is resulting in a steep decline in valuable hawks and eagles outside captivity. Many more species were vulnerable to the changing environment brought about by cultivation: for example, buffalo, bison and beavers because their habitat was destroyed, eagles and kites because they were thought to attack farm animals. Sometimes a species can revive when it is on the point of extinction – bears still survive in Spain and the lynx in Italy. The osprey has successfully recolonized Britain. But the future looks bleak for many large animals in Africa and Asia, since in general the bigger an animal is, the larger the area of habitat that has to be preserved.

Unfortunately, preserving a species is not necessarily an easy matter. Some species have become fatally dependent on a single species for food. Two well known examples are the koala bear, which can eat only eucalyptus leaves, and the giant panda, which was originally carnivorous but has adapted to dependence on an exclusive diet of bamboo shoots. If the bamboo fails to flower, the panda will die out. Other species are more versatile. Human beings are omnivores; our unspecialized eating equipment enables us to consume a wide range of foodstuffs. It is clear that for many millennia our ancestors were gatherers rather than hunters, subsisting on leaves and shoots, grubs and insects and occasional small animals.

In this first stage of their existence, the hunter/gatherer stage, human beings made very little impact on their environment. Two big changes enabled them to do much more damage. The first was the introduction of tools that made hunting for larger animals possible. Even with the primitive equipment available to them, hunters of the neolithic period seem to have been responsible for a long list of extinctions, especially in North America, where climate conditions did not favour the emergence of settled agriculture. The chroniclers tell us that Native Americans succumbed to the Spaniards because they lacked the horse; modern archaeology strongly suggests that they lacked horses because they had eaten them.

The second change came only some 10,000 years ago, with the domestication of various wild grasses: in the Middle East, wheat; in the Far East, rice; in North and Central America, maize. Abundance of food came with the ability to store dried grain, so human populations increased. Farming developed as a regular occupation, whether for peasants, farmers, stockmen or farm workers. It involved the domestication and breeding of many species of both plants and animals.

Both deliberate and accidental introduction of new species changed the nature of the biosphere. Serious damage was done, for example, to the native fauna of many tropical islands through the introduction of rats, cats and goats. The fate of the dodo on Mauritius is proverbial. On a continental scale the chain of causation can be much more complex. The British now eat onions, introduced by the Romans, and rabbits, introduced by the Normans; in turn, they were to introduce the rabbit to Australia, where it has become a major pest. Chile is one of the world's largest producers of chickens, a bird that originated in the Middle East, in what is now Iraq. The Spanish, who brought chickens to Mexico, took the turkey to Europe, where it now forms the centrepiece of a Christmas dinner, surrounded by potatoes, descendants of tubers which originally came from Peru.

The third change comes from the accelerating growth of the human population since 1500 CE. 'The progressive clearance of natural vegetation to make way for agricultural land, the elimination of "waste" ground, the draining of swamps and removal of woodland have all caused mass alteration of habitats. Ploughing of land altered the soil environment and disrupted the soil moisture conditions' (Jones, 1987, p. 71). Plants and animals have found it progressively harder to adapt to these changes, and today in the South changes are occurring so quickly that they cannot realistically be expected to adapt to them. The World Conservation Union has identified 241 centres of plant diversity, which, if protected, would ensure the survival of the majority of plant species. Most are in the tropics, in countries with limited funds for the protection of plants, and are therefore among those areas least likely to be protected.

There are three main types of ecosystems: isolated, closed and open (Jones, 1987). Each presents its own special problems, but for reasons of convenience much of the study of species has focused on isolated systems. Even there, however, scientists cannot generally say what will happen, they can only say what is most likely to happen. Ecosystems are probabilistic, not deterministic; however, the levels of probability can be high. Although such work as has been conducted on extinctions focuses on island species, the World Conservation Union identifies species introductions and loss of habitat as the two most important reasons for extinctions, followed by hunting and deliberate extermination.

Survival of species depends on two major factors: the size of the genetic pool and the habitat available. The calculation of minimum viable population (MVP) depends on the concept of 'habitat islands'; that is, just as MVPs can be calculated for real islands, so they can be calculated for remaining pockets of habitat. If real survival is to be guaranteed, moreover, the times involved are very long: 99% of the original population size has to be maintained over 1000 years.

> Mammal populations isolated on habitat islands in North American deserts for about 8000 years have provided empirical data enabling a comparison with theoretical MVPs. Agreement is fairly good ... For large mammalian herbivores 95% population persistence for 100 years requires about 1000 individuals, while for small herbivores numbers of ten times higher. The important point is that these numbers, based on demographic considerations alone, are considerably higher than those based on genetic data.
>
> (Hansson, 1992, p. 76).

However, such calculations do not take into account the very specialized requirements of many species, or the possibility of unforeseen changes in the environment, such as are likely to occur as a result of global warming. A recent requirement for major projects requiring funding by international financial institutions is an environmental impact assessment (EIA). EIAs have obvious value, and should certainly consider the impact the project may have on the local flora and fauna, even though the techniques used are still in their infancy (Jones, 1987, pp. 101–6). However, EIAs have also had many weaknesses. Those who carry them out are often themselves employed by large contractors; they are hardly likely to think differently. Even if they have been given the remit to take all possible factors into account, time and inadequate resources combine to ensure that they seldom do so. From the Southern point of view, however, the main problem is that they generally represent the views and attitudes of the North. Lastly, even when their views are taken into account, local inhabitants are not necessarily keen on biodiversity, which they may associate with backwardness and even with danger (which Indian villagers want to have tigers prowling around their house at night?), unless access to the benefits of conservation is extended to those closest to such projects.

Hence, for example, the image of Africa as a continent of game parks with numerous species of picturesque mammals is rather misleading. Parks such as the Kruger National Park in South Africa or the Serengeti National Park in Tanzania do exist, although they were originally established as game reserves and only became national parks subsequently. Since they are very popular with tourists, they generate a significant revenue, which for a poor Southern country is an important consideration. However, many areas that have been designated as 'reserves' were in the past unwanted either by the colonizers or by the local farmers, because they were very arid or because they were disease-ridden. Even then, owing to accidental or deliberate burning, or through the impact of tourism, they seldom represent a 'natural' environment (Jones, 1987, pp.141–2).

A number of African states operate a split system of conservation, preserving some sites for tourists and giving them extensive publicity, while restricting access to others which are of real scientific value. Most of these states are in South and East Africa. Even there, a growing problem in recent years has been poaching of elephant and rhinoceros. West Africa is as yet not well provided with national parks, and in some of the smaller states (e.g. Liberia, Sierra Leone) much of the variety of the indigenous wildlife has already been lost.

AGRICULTURE

All agriculture impacts on the environment in some way. The larger the scale of agriculture, the greater the impact. As with all political decisions, those favouring the development of agriculture will inevitably have unforeseen and unwanted consequences.

To begin with, settled agriculture requires that land is owned, and this in turn implies that some authority must exist to determine and guarantee ownership. It has been argued that the need to mark field boundaries in the flood plain agriculture of Egypt, Mesopotamia and China was the driving force that led to the emergence of the historical bureaucratic state (Eisenstadt, 1963, p. 4). In these states for the first time the system of government became so formalized that it continued to operate whether or not there was a competent ruler.

Not only are the historical effects of these systems still with us, so is the rural sociology of privately owned land. The tendency has long been for a small military elite to acquire big estates at the expense of their neighbours. The remainder are divided into: (a) peasants working small plots of land under traditional rights, which are often inadequate; (b) farm workers on the big estates, whose condition is often akin to servitude; and (c) people (who may also be peasants) who work part-time on the estates when their labour is needed. Since colonization was accompanied in many cases by industrialization, it is now very hard to separate out the effects of each from the other.

The industrial revolution in Europe was preceded, and made possible, by an agricultural revolution, which increased the productivity of a given amount of land dramatically. Control of political power was used to take productive land away from country-dwellers and to create large estates. In turn, the industrial revolution of the nineteenth century gave rise to transport systems which made possible the conquest of the countryside by town-dwellers. As this control has been extended in the twentieth century, it has resulted in the industrialization of agriculture (agribusiness). Similar processes have occurred in the South.

The impact on the environment of the 'internal colonization' of the countryside can be plainly seen in the replacement of traditional subsistence agriculture by large farms or plantations managed for export and/or commercial sale of a single product or limited range of products. This process was facilitated by colonization. On the pretext of efficiency, land ownership was concentrated in the hands of an elite, whose control both foreign and native owners had an interest in maintaining. Decolonization often merely replaced foreign owners by local ones enjoying the key advantage of direct access to the centres of political power. These elites have been willing to work with transnational agribusiness or be bought out by it.

The problem with agribusiness is that it tips the balance between agriculture and environment decisively in favour of agriculture. Environmental change is not taken into account in assessing the cost, nor is there any need to assess in advance the limits of sustainable land use. It is true that (in theory) more food will be produced from existing land for a considerable time into the future. However, this is not a solution, for four reasons. First, it requires high inputs of energy, which the world ecosystem is just not capable of providing. Second, it will exhaust the soil, and once that happens the situation will get steadily and irremediably worse. Third, owing to its concentration on massive production of a very limited range of plant stocks, the variety of natural plants is being lost. Fourth, in the meantime the existing levels of inequality will have to be maintained. Conflict, therefore, can arise in part from the fact that advanced industrial development is at least in part a *positional good* (Hirsch, 1977): that is, something whose desirability to others stems from the fact that only a few people at a time can have it.

BIODIVERSITY

It seems increasingly clear that biological diversity (biodiversity) is desirable in itself and may well be essential to our very survival. The consequences of the loss of the diversity of the natural world could be exceedingly serious for human beings, especially if, as is happening at present, they became excessively dependent on a very limited range of plant stocks for food. Biodiversity is not just about saving large or cuddly, furry animals — it is about the infinite possibilities in nature. From a purely pragmatic point of view, leaving aside questions of the 'rights' of other species, this could mean the difference between life and death for human beings. Most modern drugs, from aspirin to penicillin, are of natural origin. Although no one knows their number, it seems probable that among the multitude of species in the tropical rainforest there are still many to be discovered that can have important medical applications.

There are two interconnected aspects of the problem of biodiversity: first, the loss of it as a by-product of environmental damage or as a consequence of deliberate human choices, and thus the loss of ecostability; second, the commercial exploitation of it by biotechnology companies, for example. Sometimes the two occur together, as in the South Pacific area, which has the highest proportion of endemic species per unit of land area in the world and also the highest rates of deforestation. It is for the former reason that companies such as SmithKline Beecham have been investigating Fijian plant species for potential patents.

But the problem does not stop with land clearance. Large-scale agriculture by definition has worked for uniformity, not diversity. Since 1900, some 75% of the genetic diversity of the world's main agricultural crops has been lost. Livestock breeds are disappearing at an annual rate of 5%. Some 70% of marine species are fully exploited or overexploited. Freshwater fish are being lost too. But biodiversity is vital for our own food security. Uniformity means vulnerability.

There are numerous historical examples of food crop disasters resulting from a dependence on a single crop and hence on genetic uniformity. For example, the Irish potato famine of 1845–8 killed a million people in Ireland alone. The harvest failed because in warm, wet weather the fungal disease of blight destroyed the crop on which a population of more than eight million depended for food. The persistence of blight was responsible for mass emigration not only from Ireland but also from Austria, Southern Germany and Scandinavia. In Ceylon (now known as Sri Lanka), coffee, introduced as a plantation crop in the mid-1830s, was almost completely wiped out by a virulent leaf fungus in 1870. Instead, tea, which had been introduced in 1824, was planted and turned out to grow very successfully. Brown spot disease in the rice crop was the trigger for the Great Bengal Famine of 1943. Human diversity has so far ensured agricultural diversity, but that now seems threatened by the global markets that make that diversity available to others. This will mean a loss of traditional knowledge too.

BIOTECHNOLOGY AND FARMERS' RIGHTS

Maize is the product of a crossing of wild species in prehistoric times – it did not exist as such previously. There are still more varieties of maize in Guatemala than anywhere on earth, but these are rapidly disappearing. Of the five varieties of potato eaten in the Andean countries, only one was taken to North America and to Europe, becoming the basis of all modern varieties grown worldwide. Much of the responsibility for the continuing loss in agricultural diversity rests with the green revolution. In the 1960s and 1970s, well meaning breeders developed high-yielding varieties of rice and wheat, which have since replaced traditional crop varieties and wild variants in much of the South. The term 'green revolution' in itself has a political meaning. Coined in the mid-1960s, the term refers to a technological breakthrough in the productivity of crops, which, it was hoped, might be sufficient to avert the then endemic rural unrest in some parts of the South and hence to avoid or postpone the need for social and economic reform.

Productivity was increased, as high-yielding varieties of maize, wheat and rice were developed and in stages introduced to Mexico and Central America, the Philippines and later South Asia. Between 1961 and 1986, India became self-sufficient in wheat, Indonesia became a net exporter of rice and grain production increased dramatically in Latin America. At the same time, huge grain surpluses in the developed countries were available to keep down prices and fight starvation. But at the same time, the number of varieties grown dropped sharply. For example, the Philippines once grew thousands of kinds of rice, but now just two cover 98% of the rice-growing area. And recurrent virus attacks on high-yielding variety rice have already demonstrated the dangers of relying on a single strain in this way. Anxiety about the loss of genetic diversity has been increased by other problems: the high cost of pesticides and fertilizers and the pollution arising from them, the conversion of more and more land to a single crop, the high energy input required for successful cultivation and the compaction of the soil by farm machinery being only a few.

The process of trying to improve on nature has continued since. The UN World Food Summit in Rome in November 1996 followed the old formula, but under the modern guises of neoliberalism and biotechnology. It saw food security as enhanced by trade liberalization and thus large-scale, high-tech agriculture. Biotechnology research and development costs are so immense that only big companies can afford them, and they obviously want to be sure of their profits. But the high costs may mean that those who could benefit in the South cannot afford the end product, and global inequality is increased still further. For example, the Indian government could not afford the US$7.74 million asking price for the patented insect-resistant plant gene the US corporation Monsanto developed.

This inequality could be mitigated by contributions from the wealthier states. Northern countries felt that the costs of Agenda 21 and the Convention on Biological Diversity were significant, despite the fact that most such countries never contributed even the 0.7% gross domestic product (GDP) target and have been unwilling to contribute funds for *in situ* conservation of agrodiversity as advocated in the Global Plan of Action. Worse still, the saleability of all things, inherent in neoliberal ideology, has created further difficulties. In June 1996, when the FAO Conference on Plant Genetic Resources met in Leipzig, the USA sought the introduction of an amendment to the Global Plan of Action to reduce the rights of farmers collectively to continue to grow crops traditionally available to them free of charge, while increasing the rights of individual farmers to sell a patent. Here the issue of farmers' rights, which NGOs lobbied the FAO to include on the international environmental agenda in 1985, confronted changes in intellectual property rights consequent on the outcome of the Uruguay Round of the GATT and the creation of the WTO. In short, since the mid-1970s farmers in the South have found: (a) that they now have to buy new seeds afresh each year and may be legally prohibited from growing traditional varieties; (b) that these plants increasingly have desirable qualities 'built in', such as disease resistance, but that these qualities are often only achieved by growing them under carefully prescribed conditions using special pesticides or fungicides; and (c) that they therefore have become dependent in a way that they were not previously on external sources for their basic raw materials. The overall effect of this has been to concentrate rather than to disperse farm ownership and to increase farmers' debt.

Such exploitation by parts of the rich world of parts of the poor world is further illustrated by the example of the biotechnological company Phytera making deals with European plant collections by purchasing specimens for patents, without recompense to the indigenous communities and former colonial states from which the specimens were originally taken. Such communities have in recent years begun to agitate for farmers' rights, but their governments have been slow to take up their cause (*Third World Resurgence*, 72/73).

The rainforest is a particularly important concern as a rich genetic resource for plant breeding and crop improvement, not to mention the possibilities of medicines as yet unknown. But it is also a valuable commercial resource for logging and then for patenting that which remains. Many of the species within it are rare, and the limited size of their remaining habitats threatens their viability; this without us even understanding how they fit into the ecosystem of which they are part. The symbiosis of the species within the rainforest is not fully understood and many of its species are not even recorded. How could we know what possibilities it could afford the future?

CLIMATE CHANGE

We have already noted that farming is an important contributor to global climate change. The wet cultivation of rice generates large quantities of methane, as does animal husbandry. In the latter case, cutting down the trees and clearing the land for cattle ranching, as in Central America, makes matters worse; in Panama, the clearance of the forests in the 1970s has been followed by much less rainfall on the mountain ranges, so that it has been necessary at times to restrict the number of vessels passing through the locks on the Panama Canal.

In Indonesia, the very existence of our close relative, the orang-utan, is threatened by the government-sponsored rice growing schemes, which were a major cause of the burning of swamp forest in central Kalimantan in 1997, creating an acrid cloud of smoke which hung over South-East Asia for months. The fires not only destroyed the last forest refuges of the apes, but killed some of them directly and poisoned the environment for the others. It also poisoned the crops of local villagers, who as a result ate some of the apes and then trapped and sold their infants as pets, thus illustrating the way in which environmental problems have a multiplier effect for vulnerable species, *and* for vulnerable people. They also reinforce one another and, in the long term, deforestation will also change the rainfall pattern over

Kalimantan, with consequences that cannot at present be foreseen. The 1982 and 1983 episodes of burning also coincided with El Niño, destroying 20 million cubic metres of timber in primary forest and nearly twice as much in secondary forest. This loss of forest cover in turn will have helped to exacerbate the more recent drought (Hurst, 1990, p. 36).

Climate change poses an immediate threat to biodiversity. Many species are static, and even those that are mobile will be unable on current projections to move sufficiently far – either horizontally or vertically – to avoid the consequences of even a modest increase in average world temperature. But the conversion of land to agriculture also imperils members of our own species. In the 1980s drought in Africa resulted in the failure of crops in Ethiopia, the Sudan and the Sahel, followed by famine. In 1998, aid agencies were already using words like 'famine' and 'starvation' about the situation in Southern Sudan. Climate change is likely to make such situations more frequent and more severe.

NGOs AND CONSERVATION REGIMES

The different interests of North and South are illustrated by the plight of the great apes in West Africa. Like the orang-utan in Indonesia, though for different reasons, they are under threat. They are being butchered to extinction in Cameroon, for example, for the bush-meat trade. Northern NGOs, which seek to stop the trade and liberate captive chimpanzees, are thought by the local inhabitants to be trying to steal their resources. They see such policies as putting animals before people. Southern NGOs stress the role of logging for the European market in destroying the rainforest habitat of the great apes, and the loggers' willingness and economic ability to provide a lucrative market for the bush-meat trade. Again, the need to have local support for conservation measures is vital.

The issues raised in this case and also in the plight of the Indian tiger involve contests between economics and ecology and between ethnocentrism and biodiversity. They also raise important issues as to how the international community can deal with transboundary problems. Until now the only effective method has been agreement between a number of countries to create a *regime*. Regimes have been variously defined as 'a set of mutual expectations, rules and regulations, plans, organizational energies and financial commitments, which has been accepted by a group of states' (John Gerard Ruggie, quoted in

Williams, 1994, p. 27), as 'networks of rules, norms and procedures that regularise behaviour and control its effects' (Keohane and Nye, 1977, p. 19) or as 'sets of implicit or explicit principles, norms, rules, and decision-making procedures around which actors' expectations converge in a given area of international relations' (Krasner, 1983).

Some scholars argue that these only work where there is a powerful hegemon, or major power, that is both willing and able to enforce them. The earliest example, the regime for northern fur seals, was set up in 1911 by four major powers: the USA, Japan, Russia and the UK (acting on behalf of Canada). And the largest and most important agreement of this kind so far, the Antarctic Treaty, which was signed in 1959 and came into effect in 1962, involved both the USA and the Soviet Union, as well as those powers with territorial claims (Argentina, Australia, Chile, France and the UK). Young points out that the pollution control regime for the Mediterranean Basin did not include either of the superpowers, but it did include all the major states of the region that were involved on either side of any of the long-running disputes in the region: the Cold War, the Arab–Israeli conflict and the Greek–Turkish conflict (Young, 1994, p. 111).

More recently, as Young notes, 'the drive to form several of the regimes was spearheaded by intergovernmental organizations or by international nongovernmental organizations, so that states did not even take the lead in the relevant processes of regime formation' (ibid.) Examples are the International Union for the Conservation of Nature and Natural Resources (IUCN), which led to the Convention on International Trade in Endangered Species (CITES), and the United Nations Environment Programme (UNEP), which was central to the Montreal protocol on ozone depletion. The problem is that, as in the case of whaling, key states can exercise a blocking role in rendering all or part of an international agreement ineffective, and in any case it is left to the state concerned to enforce the provisions of any international regime within its own boundaries.

The case of the tiger also suggests the possibility of international NGOs being effectively bought off within a specific national context. WWF India stands accused of massive waste of funds, which are spent on administration and further fund-raising. WWF works in expensive air-conditioned offices and hotels in Delhi, dealing with politicians, bureaucrats and businessmen, while the tiger sanctuaries for which the funds were raised have yet to be established. Tiger numbers were 100,000 at the turn of the century. There are now only between 4800 and 7300 left. The tiger is mainly threatened by loss of habitat, although villagers do kill tigers when they present a threat to their villages and, more importantly, poachers kill tigers for their bones, which are used for traditional Chinese 'medicines', which have, in fact, no curative properties, and for other body parts too. It is not only the Indian tiger that is threatened. In 1998, forest fires have ravaged the habitat of the last 350 or so Siberian tigers left in the wild, causing them to break cover, where they fall victim to hunters or poachers. 'A bowl of tiger penis soup sells for up to £500 across the nearby Chinese border' (*The Times*, 28 May 1998).

The IUCN (World Conservation Union) Red List of Threatened Species published in 1996 used a new categorization of species according to the information available on them and the category of threat perceived to species by the IUCN's Species Survival Commission (SSC). It is based on the work of a number of expert committees, and is the latest and most reliable guide available on the subject. The 1996 IUCN Red List counts 30 species as extinct in the wild (EW), 5205 species as threatened (CE, EN) and 3410 species as vulnerable (VU). However, from sample evidence it is thought that anything up to 25% of the 1.8 million species known to science may be threatened; how many more species exist cannot by definition be estimated (http://www.wcmc.org.uk/data/database/rl_anml_combo.html).

CITES was concluded in Washington, DC, in 1973, and came into force in July 1975, in response to concerns that international trade was the main threat to some species. One hundred and thirty-eight countries are party to the agreement. It bans all international trade in certain species deemed to be endangered (all apes, rhinoceros, elephant, great whales, leopards and tigers), termed Appendix I, and monitors trade in other species that might become endangered or that resemble and may be confused with those that are endangered (e.g. all the Asian bears and crocodiles), termed 'Appendix II'. The Secretariat

Table 6.1 The categories of threat

Evaluated		
	Adequate data	Extinct (EX)
		Extinct in the wild (EW)
	Threatened	(Critically endangered (CE)
		(Endangered (EN)
		(Vulnerable (VU)
		Lower risk (LR)
	Data deficient (DD)	
Not evaluated (NE)		

Source: *Marwell Zoo News*, Winter/Spring 1998.

makes recommendations on scientific advice to the Conference of Parties, which meets every two years and lists species to be covered. All proposals must be passed by a vote of two-thirds of the parties. It is financed by contributions from member countries and administered internationally by a Secretariat based in Geneva. National trading is not covered by the Convention, and wildlife product, plant and live animal trading is thought to be worth US$20 billion per annum, a quarter of it illegal under CITES.

There are a number of obvious problems with CITES. Within the treaty itself there is a provision that permits can be obtained to kill endangered species for scientific research, etc. In addition, countries can declare reservations against any species when they join and any species added to Appendices I or II when they are already members; this exempts them from having to do anything about their conservation. Enforcement is left to the individual nation states. Some are very lenient with breaches of the Convention; smuggling and evasion are both serious problems. Last, but not least, many states are not members, and of those that are not members, many in the Middle and Far East trade in product(s) of at least one endangered species (Lyster, 1985, p. 9; Garner, 1996, pp. 173–6).

The Tenth CITES Conference was held in Harare, Zimbabwe, in June 1997, and the most controversial aspects were the downgrading of the statuses of the African elephant, the southern white rhinoceros and several species of whale. These downgradings represent a new challenge to the most militant environmental NGOs, such as IFAW (International Fund for Animal Welfare), which have sought to resist trade in endangered species. This challenge has become known as the 'wise use' or 'sustainable use' movement. The idea is that some limited trade in wildlife and its products gives value to the species being traded, which will ensure that it survives – the analogy often cited is that of game birds such as the partridge and pheasant in Scotland. Some NGOs have come round to this view because of the need to balance ecological against developmental arguments. The WWF is one such, and is stressing the role of eco-tourism in benefiting the people of Africa, for example. Such international organizations seek to change perceptions of local inhabitants: for example, to have the elephant recognized as a tourist money-spinner, not a trampler of crops and homes.

The Asian elephant has been on Appendix I since 1975. Numbers are thought to be in the region of 30,000–50,000. African elephants numbered 1.3 million in 1979, but the lack of control of poaching and the ivory trade led to dwindling numbers, and it was elevated from

Appendix II to Appendix I in 1989. So when CITES met in June 1997, it had been on Appendix I for eight years, and during that time the legal ivory trade had effectively ceased. Numbers had recovered somewhat, to between a quarter and a half a million by the mid-1990s. The Conference voted 76 to 21 with 20 abstentions in favour of a proposal by the host nation, together with Botswana and Namibia, to move the African elephant from 'protected' to 'monitored' status. This recognizes the growth of the elephant population and strict national anti-poaching measures in certain states, but Zimbabwe's arguments about why this should be done echo the debate between development and ecology. Zimbabwe is a poor country, which 'needs' to be able to exploit its resources, not least in order to fund further conservation. Zimbabwe's needs, it is argued, reflect its colonial history, whereby the poorest and most marginal lands were allocated to the indigenous peoples as communal tribal areas, while rich whites took the best land. The marginal areas are inadequate, especially with a growing population that cannot be expected to share their lands with growing numbers of elephant.

The new agreement, therefore, allows the international trade in ivory (from the stockpiles in Botswana, Namibia and Zimbabwe; the Zimbabwean stockpile is thought to be worth £17 million) to resume, though under strict conditions, which allow Japan to become effectively the global wholesale warehouse for this commodity. The WWF and the Environmental Investigation Agency are very concerned. They consider anti-poaching measures in Botswana, Namibia and Zimbabwe to be inadequate, and fear that legal trading will effectively conceal a flourishing illegal trade in ivory in Japan. It is certainly ironic that the world's biggest consumer should be given the formal role of regulating the trade. Further, the vote is seen as giving the wrong signal on the protection of elephants elsewhere. Since the proposal was made at end of 1996, the Indian government has reported a substantial increase in poaching and illegal ivory trading.

Other opposition comes from within the tourism industry. Tourism employs one-tenth of the world's population, and is the largest and fastest-growing industry in the world. Wildlife tourism is a major source of income for African countries, even in areas, such as The Gambia, that have virtually no large animals left. Zimbabwe earns more than US$200 million a year from tourism. In 1995, the Chairman of the National Parks Board of South Africa estimated that within only a few years tourism would earn enough to fund the entire National Programme of Reconstruction and Development.

The rhinoceros is the second largest of the land mammals and is critically threatened. Rhino horn, which is actually just compressed hair, is used to make dagger handles and an (alleged) aphrodisiac. Black rhino numbers dropped from 65,000 in 1970 to about 2400 in 1995, despite its having been put on Appendix I in 1977. The white rhino (*Ceratotherium simum*) exists in two distinct populations; both were put on Appendix I in 1975. The northern white rhino declined in Zaire (now the Republic of the Congo) from 2000 in 1970 to only 17 in 1984, before recovering slightly. Some 24 are known to have survived the 1997 civil war, but the only place where they survive in the wild is the Garamba National Park (The *Guardian*, 14 May 1998). The southern white rhino has recovered from being on the verge of extinction in South Africa in the early twentieth century to number some 7500 today. It has had a narrow escape: South Africa's proposal to resume trade in rhino horn failed to get the necessary two-thirds of the votes required by only one vote.

The theory of 'wise use' looks persuasive but the reality is that the money from trade in endangered species does not benefit those who live in the countryside, but the dealers in the market places where the products end up: in the case of ivory, Japan. These interests are heavily involved in the 'wise use' movement. So the question is: how can it be trusted? A complete ban is hard enough to police, but it is much more difficult to distinguish the products of legal and illegal trade.

FISHING

Fishing has traditionally been regarded as hunting rather than farming. The law of the sea reflects its unregulated status as commons; sadly, as a result fishermen have been consistently unwilling to face up to the inevitable consequences of overfishing. Over time, major collapses of species have become more frequent. The global marine fish catch peaked in 1989. Since then, several traditionally rich fishing grounds have collapsed from overfishing. The reason is the introduction of factory fishing methods. Large factory ships stay at sea for months, freezing their catch as they go and 'vacuuming up' whole colonies of fish by using many kilometres of fine plastic nets. These are incidentally killing many other species, notably the air-breathing dolphins and whales. Large nets are a major hazard to cetaceans; other less obvious hazards include discarded fishing gear, balls, balloons, plastic bags and other non-degradable plastic refuse. Far more whales and dolphins are now killed by swallowing such things than used to succumb to whaling.

Whales and dolphins are supposed to be protected for future generations by the International Convention on the Regulation of Whaling (ICRW), concluded in 1946. This set up the International Whaling Commission (IWC), which calculates annual sustainable quotas for each species. A complete ban is possible if stocks fall too low, but in practice whales have only been protected by the IWC once they are already on the threshold of extinction. The blue whale was not protected until 1963, when only between 200 and 2000 were left. Whaling nations seem to find it in their interests to hunt whales to extinction because of the increasingly high profits they make (Cherfas, 1988, pp. 200–3).

Several species of whale, including the giant blue whale, are now regarded as endangered, and only belatedly has strong international pressure succeeded in pressuring the IWC to impose a complete moratorium on whaling to allow stocks to recover. However, whaling for scientific purposes continues, and under the ICRW nations have only to register opposition within 90 days, and are then not bound by the decisions it makes. The threat of US sanctions did force Japan, the former USSR and Norway to accept the moratorium, and so far only one state, Iceland, has left the IWC. It did so in 1992, and plans to resume whaling commercially.

The Zimbabwe Conference received Japanese proposals to move various whales (especially the minke whale) from protected to monitored status, but did not wish to go against the recommendations of the IWC, which has zero quotas on the hunting of these whales. Japan then sought to have the control of whaling moved from the IWC, where the majority of states have little commercial interest in whaling, to CITES, where their position would stand much more chance of being accepted. At the recent conference of the IWC, Japan's position did receive support from a number of other states. Coincidentally, these were almost all small island states that had recently been the recipients of Japanese aid and had been encouraged in return to join the IWC, even though they were not active in the whaling industry and did not have the ships to be so.

As a result of the moratorium, stocks of some species, including the sperm whale, are now showing signs of recovery. The most worrying thing, however, is that the recovery has been so slight, and there is no sign that populations will or can increase to anything like the levels before the Second World War. It is even possible that the environment has been so irrevocably changed that they will never be able to do so.

For a hungry world, even greater alarm should be raised by the condition of world fish stocks. Peru used to rely on its fisheries for both food and exports of *anchoveta*, the small fish that feed larger fish further up the food chain. Industrialization of fisheries followed and the Peruvian fish catch peaked in 1970, only to collapse completely through a combination of overfishing and El Niño. More than 25 years later it still has only partly recovered, and it may be that it is unable to recover (Barkham, 1995, p. 85). If so, total world stocks of a valuable protein resource will have been permanently depleted. In other words, overfishing means a year-on-year loss of substantial food resources for the foreseeable future, and at precisely the moment at which the continuing increase in world population requires increased food production. The fact that Southern governments turn a blind eye to the consequences of overfishing for political reasons has implications that go far beyond local demand.

CONCLUSION

Even in the case of large mammals with an obvious value to tourism, and hence to the economies of the South, the North–South divide has meant that the response to the threat of extinction has been slow and often ineffectual. Although in theory it should be possible to put a price on a species that would ensure its conservation, in practice the more valuable an animal becomes the greater the incentive is to kill the remaining specimens more quickly. Conservation of rare species cannot, therefore, be left to the market.

However, biodiversity is not just about large mammals. In fact, most of the world's living species are beetles, and they play a vital part in ensuring that plant and animal life is recycled so that we can make use of it. This interconnectedness of species underpins our own existence, and as yet we do not understand it fully. The problem with the wholesale burning of the forests, therefore, is that we will not know what we have lost until it is too late. For example, it was only in 1998 that it was recognized by an ecologist from the University of Michigan that land crabs (*Gegarcinus quadratus*) play the same role in Costa Rica that worms do in Europe or termites in Africa. At night they emerge to forage for dead leaves, fruit and seedlings, and take them back to their burrows, recycling their nutrients to a depth at which they can be used by the trees of the rainforest (*The Times*, 25 May 1998). Yet thousands of acres of Costa Rican forest, home to these land crabs, have already been cleared for ranching, without any consideration of whether or not this pattern of exploitation is sustainable. Once more, the temptation has been to seek immediate returns for the local interests involved, regardless of the medium- or long-term consequences.

Reminding ourselves that development, as Gro Harlem Brundtland points out, is 'what we *all* do', we must also remember that the demand for it is the major cause of the loss of biodiversity. It is so at the level of settlers who slash and burn the rainforest, at the level of national governments that encourage such colonization or clearance for rice growing and also at the global level, where agricultural TNCs operate, supported, whether we like it or not, by all of us who buy their products.

CHAPTER 7

HUMAN ECOLOGY

Like other creatures, human beings are affected by changes in the environment: by climate change, availability of resources and the presence or absence of pollutants. The difference is that we contribute to these changes, and we also have the capacity to do something about them.

Urbanization and industrialization represent such a substantial change in human ecology that their consequences are still very much disputed. Hunter-gatherers had very little impact on the environment, and 90% of all human beings who have ever lived were hunter-gatherers. The discovery of fire some 700,000 years ago, however, marked the beginning of humanity's indiscriminate attack on the planet. Agriculture developed much more recently, but was well established in a number of places in the Middle East and China by 7000 years before the present (BP), using indigenous species. Between then and 3000 BP, population remained stable, and in consequence living conditions in these areas improved. After that time, the world's population began to increase very slowly, so that by c. 500 BP, it had roughly doubled. Even then, however, human beings lived in an essentially rural style and most people lived and died within a couple of kilometres of their birthplace. Moreover, although most low-lying land in Europe and the Middle East was cleared for agriculture, a great diversity of landscape and species survived. Only in North America, where human beings remained hunter-gatherers, were there major extinctions of other species.

A marked acceleration in the growth of population began about 1650 CE, well before industrialization began to get under way. China was as much affected as Europe. After 1750 the rise becomes exponential, accompanied by the extinction of species, the transfer of exotic species and the destruction of natural habitat.

The world population, which was only 1 billion in 1930, by 1991 had already reached 5.4 billion. It was then estimated to rise to 10 billion by 2050. This is the middle line in a set of estimates that could go as high as 12–13 billion, but may be stabilized at 8 billion. Every year, a further 87 million people are added to the world's population and the 6 billionth person will have been born early in 1999. The vast majority of this population growth is in developing countries, which include more than 90% of the world's poorest people. They not only have the least options for survival, but are most likely to degrade the environment in their efforts to stay alive. The rate of population growth worldwide has been slowing from 2.2% per annum in the 1960s to 1.5% now, but continued – albeit slower – growth is still leading to more absolute numbers. The world population grew by 40% in the twenty years between the conferences in Stockholm 1972 and Rio in 1992.

As with other species, human beings create for themselves habitats in which they feel comfortable. The UN Population Fund report *The State of World Population, 1989* pointed out that 'The earth is rapidly becoming an urban planet' (UNFPA, 1989). The traditional view was that cities were engines of development, resulting from the economies of scale they offered. A newer view is that their relative prosperity reflects disproportionate investment (Gilbert, 1976); more could be done if the same money were distributed in small and medium-sized urban settlements (Hardoy and Satterthwaite, 1986). Furthermore, there is a heavy cost. Not only are urban dwellers heavier consumers of the world's resources, but human concentrations cause heavier rates of unprocessable pollution. Cities also create new problems. They sterilize land, raise mean temperatures by some 3 °C, generate solid, liquid and gaseous waste in huge quantities, reduce run-off, but increase the risk of flooding, and create the need for elaborate transport systems and for work opportunities (Abu-Lughod and Jay, 1977; Gugler, 1996).

In 1950, most of the world's ten largest cities were in the North: New York, London, Tokyo, Paris, Shanghai, Buenos Aires, Chicago, Moscow, Calcutta, Los Angeles. Today, most are in the South: Mexico City, São Paulo, Bombay (Mumbai), Shanghai, Calcutta, Buenos Aires, Seoul. And the fastest growing cities are all in the South.

In the AICs the process of urbanization is more advanced, and they as well as upper middle-income states are now predominantly urban. Some have very high rates of urbanization indeed: Belgium and Kuwait 97%, Uruguay 90%, Argentina 88%, Germany 87%, Chile 86%, Denmark 85%, New Zealand and the United Arab Emirates 84% and the UK 83%. More strikingly, of the 42 countries listed in the high and upper middle-income categories by the World Bank in 1997, only three, Mauritius, Oman and Portugal, were less than 50% urbanized. Of course, only one achieves the ultimate level: Singapore, a special case, a city state 100% urbanized by definition.

This trend is not confined to the more economically advantaged

Table 7.1 The megacities: world's largest urban areas, 1995

City	Country	Population
Tokyo	Japan	26.96
Mexico, DF	Mexico	16.56
São Paulo	Brazil	16.53
New York	USA	16.33
Bombay	India	15.14
Shanghai	China	13.58
Los Angeles	USA	12.41
Calcutta	India	11.92
Buenos Aires	Argentina	11.80
Seoul	South Korea	11.61
Beijing	China	11.30
Osaka	Japan	10.61
Lagos	Nigeria	10.29
Rio de Janeiro	Brazil	10.18
Delhi	India	9.95
Karachi	Pakistan	9.73
Cairo	Egypt	9.69
Paris	France	9.52
Tianjin	China	9.42
Metro Manila	Philippines	9.29
Moscow	Russia	9.27
Jakarta	Indonesia	8.62
Dhaka	Bangladesh	8.55

Source: United Nations (1998), *1998–99 World Resources: a guide to the global environment*. Oxford: University Press, p. 147.

Table 7.2 The ten fastest growing major cities, 1980–1990

City	Rate of growth (% per year)
Dhaka	6.2
Lagos	5.8
Karachi	4.7
Jakarta	4.4
Bombay (Mumbai)	4.2
Istanbul	4.0
Delhi	3.9
Lima	3.9
Manila	3.0
Buenos Aires	2.5

Source: New Internationalist, No. 290, May 1997, p. 18, after UN Centre for Human Settlements (Habitat), *An Urbanizing World: Global Report on Human Settlements 1996*, Nairobi: United Nations, 1996.

states. The cities of the South are growing faster than any city in Europe did during the period of most rapid urban growth (1850–1920). And much of this growth is in primate cities, or megacities; that is, cities of more than 8m inhabitants. By the year 2000, half the world's population will reside in cities. But cities on this scale are quite new.

Urbanization has effects on both the city and the countryside. City growth is caused by the same factors that contribute to other forms of environmental degradation. *Push* factors that drive people to the city include population pressure and displacement from the land, while *pull* factors that draw them in include the demonstration effect, the availability of work and the possibilities for recreation. As Seabrook points out:

> Father Boyd says that people always give positive reasons for coming to the city; they must do this, because to do otherwise suggests that they are not in control of their lives. They say they come because of opportunities for employment, higher pay than is available in the rural areas, the chance for education, 'curiosity about life in Bangkok', the demand for labour in industry. The reality is that the force of these is easily outweighed by the negative factors: poverty, lack of land to cultivate crops, low wages; the increase in population in the countryside; natural disasters, floods and droughts; deforestation; the need to market crops at low prices because all growers bring their harvest to market at the same time; seasonal migration to supplement declining farm income. It is only later that these deeper reasons emerge.
>
> (Seabrook, 1996, pp. 25–6).

The increase in the population of cities is certainly not just a simple matter of people drifting to town in search of work. Migration takes place gradually and piecemeal; new urban residents may return to and commonly remit funds to the places they have come from. There is a link between urban poverty and rural poverty. The grey area between employment and unemployment, in marginal activity such as shining shoes, opening taxi doors or hawking, is well developed in the Southern city. The informal sector, long seen as something to be stamped out, is now seen as a source of entrepreneurial activity that is environmentally friendly, especially in its recycling aspect, and governments have been urged to encourage it (Armstrong and McGee, 1985).

Urbanization is, however, only one aspect of the increase in human numbers in the twentieth century. In biological time the explosion of the world's population is so recent that human beings have hardly had time to adapt to it. Where very rapid increase occurs in other species there are a variety of responses: migration, failure to reproduce, vulnerability to disease, etc. We have yet to see the full consequences of crowding human beings into cities. But the growth of cities clearly made possible the rapid spread of epidemic disease from ancient times at least down to the great influenza epidemic of 1919.

It is of course generally agreed that human beings do not act because they are driven by purely biological considerations. However, there is also evidence that increased population leads to overcrowding and hence to organizational instability. Urbanization therefore creates stresses which are likely to be released unpredictably (Leroy, 1978, pp. 71–2).

Cities grow because they attract migrants. West Africa is one of the poorest parts of the world; it also has one of the highest rates of urbanization. In West Africa generally, much of the nation's manufacturing industry is to be found in the capital city; in addition, the central government has tended to concentrate its investment there. Three strategies have been used to try to resist this situation. The most drastic is to move the national capital, as in the case of Nigeria from Lagos to Abuja. Nigeria, Côte d'Ivoire and Senegal have also tried the second approach: to try to develop a number of regional centres. Another strategy has been simply to try to stop migrants from coming in. However, none of these has been particularly successful (Salau, 1990).

Penang in Malaysia, a relatively small island, has been transformed out of all recognition:

> Almost the entire island of Penang in Malaysia is now semi-urban: tourism, free trade zones, housing for expatriates, golf resorts, road and bridge construction have swept away the old villages and imposed upon Penang the aspect of a sprawling townscape which has drowned the elegant old colonial capital of Georgetown. Add to this the migration of those squeezed out of farming, or displaced from old rubber plantations, people who have no option but to come and try and find a place in the free trade zones or the service sector of Penang, and it is easy to see how the lineaments of the island culture, fishing, *padi* and trading were so quickly effaced.
>
> (Seabrook, 1996, p. 165).

Alongside their sterilization and pollution of the surrounding countryside, cities provide a sharp contrast with the rural hinterland where there are very few services, and this lends them a spurious glamour. In many parts of the South, the absence of mains electricity means that farmers (and especially their wives and children) work hard and are relatively deprived compared with the urban population. But

Southern cities are not always very advantaged: even before the recent civil war, a chronic shortage of generating capacity meant that many parts of Freetown, Sierra Leone, were without electricity for much of the time, and also without refrigeration.

The rapid growth of cities, too, may mean that water supplies are inadequate or of poor quality. Even in relatively prosperous Southern countries, housing can be very basic and mains services scarce or non-existent.

A survey of Bridgetown, Barbados, shows that 57.3% of houses are built of wood; a figure which goes up to over 90% in at least one inner urban area. The overall national figure for pit latrines is 52.2%, and this figure is exceeded in half of the districts of the capital, often by a considerable amount. This situation is explained historically by the tradition of the 'chattel house' built on poor land at the edge of a plantation. This insecurity of tenure had an advantage, in that it made it easy to move, but also a negative consequence, in that there was no incentive to improve the standard of housing. Although in no way short of self-help, Barbados, a relatively well developed country, has as a result a very poor standard of urban housing (Potter, 1990).

It is the pollution cities generate that is the thing visitors notice first. In 1992, WHO and UNEP jointly published a report on air pollution in 20 of the 24 of the world's megacities (here urban areas with more than 10 million people then or by the year 2000). Every one of the cities studied had at least one major air pollutant that exceeded WHO guidelines, 14 had at least two, seven had at least three. Mexico City has the worst overall air pollution.

On the other hand, there is a strong argument that it is environmentally desirable for human beings to live in cities. Not only can they enjoy all the advantages of social existence, but efficient arrangements can be made to dispose of waste and so on, and the blight of the countryside can be arrested. At least, this is possible for some of those who live in cities.

Shanty towns are the most obviously distinctive feature of Southern cities (Gilbert, 1994). They are as much a feature of urban life in Africa as in Latin America (Simon, 1992) or South and South-East Asia (Seabrook, 1996). Technically, those who live in them have no formal rights to the space they occupy, and they are therefore commonly referred to in English as 'squatters'. In the South, however, the term 'squatting' involves illegal occupation of land rather than of the building. There are three main processes by which land is converted to squatters' use: mass invasion, pirate subdivision and infiltration (McAuslan, 1985). Main (1990) points out, however, that many squatters in Kano, Nigeria, took to squatting because in the first place they were dispossessed of land they held on customary tenure. The 1978 Land Use Decree enabled the state to expropriate land more easily. 'Although its powers of expropriation and reallocation might in theory be used for equitable ends, they have in fact been used quite consistently to benefit those with sufficient wealth or political power, or with the right contacts' (Main, 1990, p. 20). In Kano, in theory tenure can be obtained by squatters, but few understand or can make use of the processes to obtain it. However, it is also true that settlement that is 'illegal' in the formal sense, i.e. has not taken place according to the forms established by the Northern law implanted in the colonial period or according to laws made by the urban elite since, may be – and often is – 'perfectly legitimate' in West Africa according to customary law (Simon, 1992, p. 109).

The problem is not just that the rules have been changed, but that they have been changed according to a view of the world in which the market is expected to satisfy the need for housing that will enable ambitious governmental plans for industrialization to be maintained. However, in very few of the large Latin American cities, which began to industrialize earlier than those of Africa, was there any systematic attempt to build housing for the workers. Buenos Aires, São Paulo and Monterrey are exceptions; on the other hand, Brasília, despite being designed by a world-famous communist architect, is not. As they had nowhere else to live, therefore, the in-migrant workers occupied any land they could find, mostly on the edges of the cities, but some in the central city area, as on the hillsides of Rio de Janeiro or in the ravines of Guatemala City (de Oliveira and Roberts, 1996).

In Mexico, as in other Latin American countries, squatters used automatically to acquire rights to land if not challenged within five years, but in 1992 the law was changed to end this. Hence shanty towns are a feature of all large cities in the region. As Wayne Cornelius warns us, however, they are very varied, and although they are often the first place of residence for in-migrants to the city, they also contain a great number of native-born urban dwellers (Cornelius, 1971). In Rio de Janeiro, the settlement of Rocinha, on the hillside above Copacabana, has been in existence since the 1950s. It was and is still viewed with great distrust by the well-to-do. However successive governments wisely left the *favelas* of Rio to their own devices until, on the weekend of 18–20 November 1994, ostensibly in a crack-down on organised drug-trafficking, teams of soldiers were sent into Mangueira and other smaller shanty towns. They encountered no significant resistance (*Keesing's*, 40274).

Past history explains this to some extent. In the 1970s, the cities of Latin America were used as a battleground for urban terrorists bent on seizing political power. These movements were suppressed, in some cases very ruthlessly, by military dictatorships. The military governments of the period, however, spent money unwisely and so helped to precipitate the debt crisis of the 1980s, one of the consequences of which has been a steady decline in the quality of the urban environment, as funds were diverted to other purposes (de Oliveira and Roberts, 1996). However, if today the shanty towns are stable enough socially and politically, from an environmental point of view the situation is very different. For example, in the marshy Baixada Fluminense north of Rio, the watercourses run with effluent and disease is rife. Although the hillside settlements are healthier, they have been known to collapse into the ravines during the heavy rains.

The colonial city in Africa can be regarded as a kind of pre-industrial city characterized by small-scale production. However, the argument that colonial cities' distinguishing feature is their role as an 'alien implant' in the local culture is compelling, given the way in which buildings and layout reflected the practice of the colonizers rather than the colonized. Manuel Castells (1977) writes of 'dependent urbanization', in that the cities formed a key economic link between the colony and the global economy. The colonial port cities of Africa have not in general remained major centres of economic activity by world standards since decolonization. However, cities have become the favoured residence of an urbanized elite, and the gap between town and countryside in social terms has, paradoxically, actually increased in most countries since independence. Dakar, Senegal, where a city of some 1.5 million inhabitants has expanded to fill almost the whole of the Cap Vert peninsula, is a good example, with high-rise blocks housing the diplomatic and commercial elite and the workers in slums on the outskirts (Simon, 1992, pp. 171–3, see also Arecchi, 1985).

This is the more surprising, since the other most striking feature of African cities is that since independence controls on migration and residence have been relaxed. 'Irregular land occupation and housing have today become ubiquitous characteristics of African cities, particularly since the easing of repressive measures and migration controls at the time of decolonisation' (Simon, 1992, pp. 106–7). The causes of this include the survival of customary tenure alongside formal legal tenure, the rapid rate of urban growth, inadequate and/or inconsistent planning procedures and shortage of funds. Formal controls have seldom been consistently applied for long, and from the 1970s on there has been a general tendency to abandon any attempt to replan the city as a whole and to allow the upgrading of irregular housing. In Addis Ababa and Mogadishu, over 80% of the cities' populations live in irregular housing, and in many other major cities the figure is also very high.

The colonial port cities of Asia retain their property of acting as magnets of development. Bombay (Mumbai), Calcutta, Colombo, Penang and Jakarta were all situated where they were to facilitate access from the sea, making use where possible of rivers and lagoons. They were unhealthy places in colonial times, and even today flooding during the monsoon season remains a chronic problem for the modern residents (Murphey, 1996).

On the outskirts of Dhaka, capital of Bangladesh, the newly arrived are crowded on to land beside a railway line and as close as two metres to the track.

> The moon provides the brightest light: on both sides of the track the huddled outlines of hutments are visible, some of them semicircular, flexible bamboo draped with polythene; it is a camp rather than a settlement, a transit camp for those displaced from country to city. The huts are all on low ground; when it rains the embankment is weakened and the floodwater simply cascades into them. Here and there is a brighter light: the fluorescent strip of a pharmacy, where antibiotics, painkillers and remedies for jaundice and dysentery absorb so much painfully earned money.
>
> (Seabrook, 1996, p. 128).

Interestingly, since independence, the primacy of Jakarta in Indonesia has grown. Although it still has only a small percentage of the country's population, it is now clearly its leading centre of economic activity, with the obvious consequences for the environment:

> Greater attention needs to be paid to the environmental consequences of Jakarta's growth. For example, the excessive drawing of water from underground aquifers has lowered the water-table causing an encroachment of seawater into much of the aquifer and causing building subsidence. Air pollution is at levels above WHO standards in some areas. Environmental health considerations must be recognised and given legislative and financial back-up.
>
> (Hugo, 1996, p. 177).

However, they will not be. Where politics is dominated by a fairly small elite — as was the case in Indonesia for more than three decades under President Suharto and his relatives — they have the incentive and the opportunity to escape the consequences of the activities which generate the wealth on which they rely. They have already spoilt much of the charm of colonial Bandung:

The main shopping street, Jalan Braga, which once aspired to rival the Champs Elysées, has lost its elegance under a mask of aluminium and dirty concrete. But for Asia's crisis, it would now be sprouting a large shopping complex, part of a project backed by 'Tutut', President Suharto's eldest daughter, who is the welfare minister and a millionaire businesswoman.

(The *Guardian*, 3 April 1998).

The problem of municipal waste is compounded by the dependence of many Southern countries on the income from tourism. Tourists expect Northern facilities in Southern states, and to give them what they want in the shortest possible time there is a strong tendency to present them with a standardized product, much of which may be 'disposable' (as in the case of airline meals, for example). The result is that tourists generate quite disproportionate amounts of waste products, both organic and inorganic.

The impact of the growing demand for leisure facilities on the environment begins with the 'urbanization' of the land by building hotels and leisure facilities. These in turn put a heavy demand on public services such as power, water supply and sewage. In the poorer states with limited public facilities, tourist hotels can be built with totally self-contained facilities, powered by their own generators; an island of the First World in the middle of the Third. Whether or not this does anything for the tourists' perception of what life is really like in the countries concerned is, of course, hard to judge, but it does seem rather unlikely.

Transport and its inadequacies form an additional set of problems. The economics of housing and transport are closely interlinked. The majority of jobs in the formal sector in Bombay, for example, lie within two miles of the city centre. The result is that the railway and road network is scarcely able to cope with the commuter traffic. The Mexican government, calculating that the traffic jams created by the capital's 2.6 million cars were costing 1.3 million person hours a day in productivity, built a vast metro system and banned the use of cars on one day in every two. Traffic in Seoul, South Korea, is so congested that a city centre bridge collapsed in 1996 under the weight. In Cairo, traffic lights are ignored as having no relevance to the real needs of the situation.

In addition, of course, large quantities of food have to be brought into the megacity each day to feed its swollen population. Any problem in this regard and the city authorities could find themselves having to cope with riots.

. Apart from the growth of shanty towns, the subdivision of existing housing reflects the desire to be as near the possibility of work as possible. But as congestion increases, the social structure of the city breaks down. The persistent problem of all developing countries has been underemployment; hence the importance in, for example, Indonesia of small-scale activities in the informal sector 'without legal permits, with little capital and low earnings per worker' (Forbes, 1984, p. 167). Corruption and criminality, are, therefore, the inevitable accompaniment of rapid urbanization; this presents a problem for agencies such as the World Bank, which recognize the importance of the informal sector but find difficulty in devising strategies to target it. There is in the meantime a tendency to play down its negative aspects.

New arrivals in the city retain links with their place of origin, but swiftly integrate in new communities. The problem is that they seldom have the skills to do more than survive in the new conditions. Community politics is the key to empowerment, and new social groupings have emerged in some countries as the vehicle of modernization.

LAND USE AND SETTLEMENT

Increases in the 1970s in the land area cultivated in Latin America, China and South and South-East Asia have been considerable, though at the cost of damage to marginal land and its fragile ecosystems. The 'carrying capacity' of the land – that is, the number of people a given area can support – has been increased at least temporarily, as a result of the 'green revolution', involving the use of high-yielding strains and chemical fertilizers. On the other hand, there has been a marked decline in the production of both wheat and maize in Africa, despite (or because of) the increase in production of cash crops for export, such as tobacco and cotton.

In India, the green revolution brought only temporary benefit. In the 1960s, India was purchasing up to one-fifth of the entire US wheat crop to feed its people. Within a decade it had become a net exporter of grain. However, with a rapidly rising population, productivity was actually less per capita than it had been at the beginning of the century, under colonial rule. In 1992, imports of grain were resumed. In the words of Lester Brown, founder of the Worldwatch Institute:

> At the time we knew what needed to be done in countries like India ... We had to change pricing policy and get fertilizer there and new rice and wheat seeds that we knew would work because they were tested. All the ingredients were there. But now that India and other developing countries have made these changes, we don't know what to do for an encore.
>
> (Moffett, 1994, p. 71).

The impact of the green revolution has been felt in three ways: land clearance, the exploitation of marginal land and the expansion of large landholdings growing cash crops. Land clearance has resulted both in damage to the land from the process of clearance itself, opening the soil up to erosion, and in inexperienced hands has led very rapidly to the degradation of the land and to its abandonment. Marginal land is by definition problematic. It requires careful treatment and in any event will never be as productive as the better quality land. But ironically it is marginal land that most Southern countries rely on to produce the bulk of the food consumed locally.

The reason is that the power structure has consistently been used both by local elites and large, often foreign-owned corporations, to create large estates. These estates, however, can only compete on the world market in specific commodities. Much of the best land in the South, therefore, is monopolized by relatively few owners and dedicated to large estates growing the cash crops that generate really good returns in the export market. However such commodities do not contribute in any way to a sustainable pattern for the local inhabitants. Instead they become dependent on the fickle effects of international supply and demand for products of relatively little food value such as coffee or sugar or wholly inedible products such as cotton, tobacco or rubber, for which local demand is finite.

Estates vary a great deal. They have existed in their present form in Latin America since the sixteenth century, and Brazil has experienced three successive 'boom' periods, in sugar, rubber and coffee respectively. In South Asia, plantations have been established since the nineteenth century, producing tea and cotton. In tropical Africa, plantation agriculture often dates only from the beginning of the twentieth century, and crops produced include cocoa, peanuts and palm oil. Of course, a considerable part of plantation production is consumed within the South (for example, before the Gulf War, Iraq was the major consumer of Sri Lankan tea). Nor are all plantation crops necessarily in direct competition with local food production. Tree and shrub crops such as tea and coffee usually occupy a relatively small part of a country's cultivated land. Even bananas, which as a cultivated crop are very wasteful of land, are largely grown for consumption within the South.

In recent years, drastic industrialization of the countryside has not only done widespread ecological damage, but sent great waves of immigrants into the already swollen towns. 'One result of this', writes Jeremy Seabrook, 'is that the rural poor must feed the city with their children. The small farmers, in any case, are cheated by middlemen, dealers and officials; which is why so many have almost come to believe that the production of food, far from being the most useful and valuable work of humanity, is actually an ignoble, inferior occupation, from which they want their children to escape' (Seabrook, 1996, pp. 28–9).

LAND REFORM AND LAND COLONIZATION

One of the push factors contributing to urbanization is that, with rising populations, too many people in the South are working too little land to be able to support themselves by that alone. There are two possible 'solutions': land reform by redistribution, which is politically difficult with the vested interests involved; and land colonization. The latter is of course only possible where land is available. The former is only possible where the elite that controls the land is prepared to let it happen.

In Guatemala, for example, a modest attempt at land reform occurred in the early 1950s. However, soon afterwards the government of Jacobo Arbenz was overthrown by an insurrection sponsored by the USA, and the land reform was largely reversed, while political control was maintained by a series of military governments fully prepared to use force to maintain the status quo. In the 1970s, they turned instead to encouraging people to migrate to the northern province, El Petén, cutting new farms out of the virgin rainforest. Once it appeared that there might be oil under part of this area, the armed forces moved in, carving out estates for themselves. In under 25 years Guatemala lost half its forest cover, while the race war that has raged since 1978 came to an end in an uneasy peace only in 1997.

Land reform involves the breaking up of large estates and the distribution of plots to individuals or groups of individuals to work them. Inevitably there are problems with this process, even supposing a government can be found that is prepared to do it. Apart from the former communist countries and Cuba, relatively few countries have carried out effective land reform except under military threat, but the south-west Indian state of Kerala is an example of such a successful land reform. Even if the principle is accepted, there are clashes of interest between three groups: those who work the land; those who live nearby and wish to use the land; and those who live in urban areas and want to have access to land. Another problem of land reform is that productivity usually drops sharply immediately afterwards, as in

Cuba in the 1960s, as people adapt to the new order of things. Once sustainable productive patterns have been established, however, there is no reason to suppose that this decline must continue.

The easier option, where it exists, is land colonization. Land colonization has a detrimental effect on the land where, as it usually does, it results in permanent land clearance. Initially the methods are very similar to that of traditional 'slash-and-burn' agriculture. This is still practised, usually in tropical rainforest, sometimes in savannah. The main locations which have been significant in recent years have been Central America, western Amazonia, tropical Africa, the Philippines, Malaysia and Indonesia. A fresh site is selected, trees are cut down, leaving larger tree stumps in place, and then branches, twigs and bushes are burnt, leaving the charred landscape covered with a layer of wood ash, which acts as a fertilizer. Intercropping of species is normal and the system, which was devised for local crops, such as manioc and cassava, has successfully assimilated crops introduced from elsewhere.

When this is intensively practised the results are very different. 'Slash-and-burn' cultivation can be used successfully over very long periods, provided not too much is asked of the land. But the fertility of the thin forest soils falls off rapidly, and yields in the third year are normally only half that of the first. The success of the method therefore depends on the cultivation of relatively small patches which are shifted every two or three years. One or two years of crops in Brazilian, West African or South East Asian rainforest is followed by periods of fallow varying between eight and fifteen years, during which secondary vegetation is re-established and the forest begins to regenerate. It is impossible to accelerate this cycle without damage, but this is in effect what new settlers are trying to do. Land clearance opens up areas to permanent settlement. Population growth leads to a reduction of fallow periods. Global demand leads to the sale of the land to large landowners or corporations (or its usurpation). This is followed by intensive cultivation by mechanical and chemical means, and attempts to realize increased profit, in particular by the introduction of ranching after the initial decline in fertility. The outcome is the permanent degradation of the land, resulting over time in a general decline of its carrying capacity in the face of rising demand.

Paul Harrison gives the example of the village of Ambodiaviavy, near Ranomafana in Madagascar (Harrison, 1992). This was carved out of the forest in 1947 by eight families, numbering in all 32 people, who settled on the valley bottom. Between 1947 and 1980, a combination of the natural growth of population and the arrival of new settlers from the overcrowded plateaux raised the population of the village to 320. With the best land already occupied in the 1950s, cultivation gradually spread up the hills on both sides of the valley. However at the same time existing family holdings were subdivided to give new plots to children at marriage, so most of the plots of the 36 families resident in 1980 were too small to feed the family. Families responded in a variety of ways. Some sold part of their plot, making their situation worse rather than better. Others tried to carve out new plots from the steep forest land higher up the mountain. Not only was this land relatively unproductive, but in a short time their need for food forced them to cut down on the vital fallow period needed for the land to recuperate, with the result that its capacity to regenerate was being permanently impaired. In a single generation the families had gone from relative affluence to poverty.

Madagascar is a big enough island still to be able in theory to feed all its inhabitants. However, the FAO has identified a number of 'critical zones' worldwide where land resources were already inadequate to feed their 1975 populations. Most of these were areas subject to severe land degradation, where the natural carrying capacity of the land was already seriously impaired, and their total population was in excess of a billion (Higgins *et al.*, 1982).

Where land reform is not carried out, or is inadequate, new problems arise as would-be farmers try to carry out organized land occupations. In Brazil, the 1991 decree prohibiting non-Indians from appealing against the decisions of the Indian Affairs Bureau (FUNAI) on the demarcation of tribal lands was significantly eroded by new legislation promulgated by Presidential decree on 8 January 1996. The Catholic Church's agency for Indian affairs denounced this as an attack on the constitutional rights of Brazil's indigenous peoples. A leader of the Landless Peasant Movement (MST), Deolinda Alves de Souza, was arrested on conspiracy charges on 25 January but released in March. However, it was only after the massacre on 17 April of between 19 and 23 peasants by police bent on clearing an MST road blockade at Eldorado de Carajás, state of Pará, had aroused widespread protest that the government addressed the land question directly. Asking Congress to pass legislation for the expropriation of uncultivated land, President Fernando Henrique Cardoso gave the National Institute for Colonization and Agrarian Reform (INCRA) Cabinet status, and appointed Raúl Jungman of the centre-left Popular Socialist Party (PPS) Minister for Land Reform. José Eduardo de Andrade Vieira, who had refused to condemn the police violence, was dismissed as Minister of Agriculture, and his post was given to Arlindo Porto of the centre-left Brazilian

Labour Party (PTB). The MST, however, dismissed these moves as cosmetic and resolved to continue the land occupations that had already taken place on 168 estates. A further 12,000 people occupied 83,000 hectares at Rio Bonito do Iguaçu, state of Paraná, on 19 April. Meanwhile, the main result of the policy of encouraging settlement in the Amazon basin has been to delay, without preventing, the eventual arrival of immigrants from the North-East in the big cities of the South.

POPULATION

Poverty and population growth are clearly related. However. the nature of that relationship, and even the direction in which it operates, is still debated. Does population growth cause poverty or poverty stimulate population growth? The Prince of Wales has suggested: 'we will not slow the birth rate much until we find ways of addressing poverty; and we will not protect the environment until we address the issues of population growth and poverty in the same breath' (Prins, 1993). This to some extent seeks to bridge the gaps between the two main groups in the debate: the neomalthusians and the neoliberals.

Neomalthusians argue that finite resources will be outstripped by rapidly rising population, leading to ecological overload and mass starvation (Ehrlich, 1968); population increase is the motor that drives poverty. This argument has been used to support coercive population policies, such as the Indian sterilization programme. Most of this growth is in developing countries. However, what is not always noted is that Ehrlich also argued that a major problem was overconsumption in the USA (Moffett, 1994, p. 110). The North has 20% of the world's population but uses 80% of its resources, so there is not enough left over for everyone else. 'Both poverty and wealth degrade the environment in different ways' (Thomas, 1992, p. 10).

Neoliberals say that the free market will ensure that there is more than enough food; human beings will continue to be able to provide for themselves through technological advances. Population growth multiplies economic activity. 'More people not only means the use of more resources but more units of creativity and productivity. More people compete creatively for ways to develop or find substitutes. Thus the world's resources are not finite' (Simon, 1980). So, either way, population is not the only issue. As Porter and Brown (1991, p. 2) put it:

> It is not the growth of population per se but the total world population multiplied by per capita consumption that is the important measure of total potential stress on the global environment. The per capita *gross world*

product (GWP) – the total of goods and services produced and consumed per person throughout the planet – has been growing faster than world population for decades.

However, the world is not one in political terms and the link between population and sustainability cannot be so easily dismissed. Norman Myers writes of the 'collision between human numbers and the resources needed to sustain them' (Myers, 1991, p. 5), and the fact is that over the past 25 years humankind has been depleting natural resources such as land, water and energy at an accelerating rate. Appearances can be deceptive. The world is a big place, and the expected population growth of some 3.8% per annum in Kenya, Zambia, Tanzania and Côte d'Ivoire does not seem much in such a huge and relatively empty region, but this rate of growth will mean that these countries double their populations inside 20 years. They are just the most obvious examples in a region of rapidly growing population and very limited resources (Tobin, 1994, p. 277).

It is true that, while consideration of consumption cannot be left out of the equation, population growth does act as a 'multiplier of economic activity'. Despite its high rate of population growth, Africa is not – yet – a densely populated continent. The most densely populated parts of the world are in Asia and Europe, but how far given populations can be sustained clearly varies, reflecting the unequal distribution of the world's resources. Population growth combined with poverty in areas vulnerable to natural hazards but seeking development is a potent force for environmental degradation. No one knows what the 'carrying capacity' of the Earth is, but it is certain that all our problems, especially environmental ones, will be greater with a larger population.

In this debate, the first thing to be clear about is that estimates of future population growth have to be taken seriously. Early attempts to project population growth were relatively simple, seeking only to estimate what the future population of the world and its principal regions might be. But the first UN projection for the year 2000, prepared in 1957, estimated a global population ranging between 4.9 and 6.9 billion (United Nations, 1958). If, as now seems likely, the actual figure will be 6.3 billion, this suggests that we can have some confidence in current projections at least up to the year 2030 (Frjeka, 1994) and would be wise to be pessimistic.

More recent projects have sought to elaborate a number of different scenarios for future growth, allowing for different assumptions about fertility, mortality and migration. These are intended to give clearer guidance to policy-makers as to what the consequences are likely to be of different policy alternatives. The problem is that they also offer ample opportunity for politicians to evade having to make hard choices. Furthermore, although the data on which the projections are based are more ample and more detailed than ever before, most of the projections that have been made have been based on the assumption that the end point will be stationary populations with a reproduction rate of unity, and this theoretical steady state cannot be taken for granted (Lutz, 1994, pp. 11–12).

An International Institute for Applied Systems Analysis (IIASA) study, which takes into account the possible influence of environmental changes on fertility and so on, shows clearly that substantial further world population growth is certain. 'Unless an unpredictable, major disaster occurs, the world population will grow by at least another 50 percent to above 8 billion before the year 2030. It may also increase by 100 percent and reach close to 11 billion by 2030.' The central projection is for a global population of 9.4 billion (Lutz, 1994, p. 403). However,

> Trends in population growth are extremely divergent in different regions of the world. In Western Europe, the central scenario yields a mere 10 percent increase by 2030, but in sub-Saharan Africa it results in a tripling of the population. The central scenario results in a 35 percent increase in North America, a 75 percent increase in South America, and a 50 percent increase in China. Roughly a doubling of the population will occur in North Africa and South Asia as well as in Central America and the Caribbean.
>
> (Lutz, 1994, p. 405).

Hence, by 2030, between 85 and 87% of the world's population will live in what is now the South. The mean age of the population, which has already risen in the AICs, will also rise elsewhere. If fertility is low, the ageing will be very marked: the world mean will certainly rise to 31 years and could rise to 35, the present mean age in North America. This means that proportionately there will be many more women than men in the older age groups.

There have, of course, been a number of attempts to calculate the world's 'carrying capacity', using the best data available about soils and other features and making various estimates of the extent to which artificial intervention will continue to increase productivity. Recent estimates range from two billion to one trillion (1000 billion), although this last figure is so far in excess of any other estimate that, like Simon's contention that there is no limit, it can safely be disregarded as irrelevant. In theory, using current technology, Sudan could produce enough food for the whole of Africa; in practice it has been racked with famine. 'Most experts would probably agree that we could sustainably supply the present 5.5 billion world population' (Gerhard K. Heilig, in Lutz, 1994, p. 253). However, although Heilig thinks a population of 10 or 15 billion could be sustained, this would only be possible if a formidable list of conditions were met in the next few decades.

The most recent trends suggest that even this amount of optimism may be misplaced. World grain production per capita rose steeply between 1950 and 1984. Since then it has fallen by about 10%. Meat production per head has continued to rise, but only through the input of large quantities of feed grains; beef, which depends on the availability of grazing land, has been in decline since the early 1990s. Most alarming of all, as we have already seen (Chapter 6), fish stocks in several of the major world regions have collapsed, and there is little sign of concerted action to save the rest (Brown and Kane, 1995).

Part of the problem is that the world's carrying capacity is most unevenly spread. Africa has 20.2% of world's land area and only 14.9% of the world's population. In parts of Africa, such as the Republic of the Congo, land is abundant and, with inputs, would be capable of feeding many times its present population. Poverty, conflict and corruption have made other states — Kenya, for example — quite the opposite. It is not only the depletion of resources through higher human numbers but also the pollution factor. The WHO submission to UNCED stressed the need to control population to spare deaths consequent on environmental pollution (WHO, 1992).

People everywhere have aspirations to goods such as fridges and cars. But the need for food, as land gets scarcer, leads to increased intensity of agriculture, including the greater use of pesticides and fertilizers and the growth of urban areas with displacements from the land. Developing nations at UNCED were unhappy with the interjection of the population question, which they saw as designed to distract attention from Northern consumption patterns. Hence, the South has not yet taken the problem of pollution resulting from increased population and thus from increased economic activity seriously, and there are likely to be increasing problems from this source.

LIMITING POPULATION

The ways in which population increase interrelates with other aspects of the environment are well illustrated by the case of Egypt.

Population increase threatens the security of a country like Egypt because it puts pressure on the resources the country requires, most notably, in Egypt's case, the waters of the Nile. With eight countries upstream sharing the river, Egypt was in a vulnerable position to begin with. Population growth has increased its vulnerability. Part of the problem is the geometric progression of population growth correctly foretold by Malthus. In a country like Egypt, where more than 50% of the population are under 15, this is especially marked. But there is also the legacy of old attitudes, especially in rural areas, where literacy is less than 50% and large families are still the norm. Increasing congestion and large-scale unemployment fuels bitterness, which has taken the local form of Islamic fundamentalist terrorism. Yet Egypt is in fact a success story in the voluntary limiting of population growth. In only ten years, through education and campaigns to change attitudes, the average family size is down from seven children to four.

Only six countries were able to reduce their rate of population growth to below 2% per annum by 1980–5 and to cut total fertility rates by 30% or more since 1960. These states were China, Sri Lanka, Colombia, Chile, Burma and Cuba. There is no one set of factors that explains all these cases. In only three of them was there a strong public effort to limit population growth. However, two other factors do seem to be relevant: relative food security and an enhanced status for women (Lappé and Schurman, 1989, pp. 55–63). Two NICs, South Korea and Thailand, have also been very successful in restraining population growth – the latter reduced its growth rate by half in 15 years by popularizing contraception and making it seem fun (Tobin, 1994, p. 280). Note also that Thailand is a country in which women have more equality and freedom in their personal lives than in many others.

The role of women seems to be the most important factor in Colombia, where seven children was the average 30 years ago, compared with 2.8 now. There, a non-profit-making organization relying on those it serves to pay for its services has had tremendous success. PROFAMILIA has not had any help from the Colombian government, but its aims have been in accord with the government's perception of what is best for Colombia, and as a result its work has not been obstructed. In Costa Rica, the availability of a free public health service has obviously been a key factor.

The government of the Philippines, like that of Colombia, recognizes population control as desirable, though with a annual growth rate of 2.4% and a more limited resource base, it is prepared to go further in actively supporting family planning NGOs like the

Gabriela Foundation. It also finances field clinics and the giving of contraceptive advice to women in hospital after childbirth. In so doing, it has incurred the wrath of the Catholic Church. In Bangladesh, as many as 700 private voluntary organizations are registered. A US Aid programme was begun in 1982 to educate girls, but since 1973 it has been barred by Congress from funding abortion services. Instead, the Bangladesh Women's Health Coalition (BWHC) runs rural clinics and provides reproductive health services. Middle-class homemakers linked health and control of fertility in Concerned Women for Family Planning when working with the displaced people from the famine of 1974 (Moffett, 1994, pp. 197, 199, 201). Both the UN Conference on Women, held in Nairobi in 1985, and the UN Conference on Human Rights, held in Vienna in 1993, endorsed their efforts.

These examples are of voluntary schemes quite unlike the coercive measures adopted in the Emergency in India in the 1970s, or the One Child policy operative (for some) in China. India's programme began in the 1950s with modest and rather half-hearted measures, such as the raising of the minimum age of marriage and recommendations about the education of young people. The limited success achieved through these schemes led several states to opt for coercive measures, such as loss of posts for public employees with two children who refused sterilization. This was taken further still by overseer Sanjay Gandhi's flying squads, but this was fortunately short-lived. The programme's overall success was to reduce India's fertility rate by some 30% between 1960 and 1990, but it is unlikely that this success will continue.

China reduced population growth from 2.2% in 1970 to 1.1% in 1985, but resistance and consequent exemptions, especially among the 80% who live in rural areas, where families need child labour, led to a substantial increase in rates of growth in the late 1980s, and China abandoned its rather ambitious population goals (but see also Qu and Li, 1994). Thailand and Sri Lanka have achieved similar levels of reduction to China without the violation of human rights that coercive policies involve. Jordan, Kenya and Bangladesh are other examples of greatly reduced family sizes, which suggest that the grimmest estimates of population growth may be avoided.

These successes are directly due to the use of artificial contraception. In 1965, 10% of couples in the developing world used contraceptives and the average family size was just over six children. Now 55% use contraception, and the average family size is just under four children. But the willingness to limit family size correlates very closely with female literacy. It is clear that prosperity (and especially female education) slows population growth. Poverty places a high premium on large families, both for work (especially in rural areas) and for insurance in old age. Affluence makes the choice of smaller families possible. At the Cairo conference in 1996, there was a reluctant recognition of people's right to limit their families if they wish, and an important reason why they do so is to escape from the economic burden of child-rearing. But there was also a very powerful lobbying effort by the Catholic Church, in an improbable alliance with Muslim fundamentalists, to prevent the endorsement of artificial means to contraception.

The striking relationship between education and slowing population growth has encouraged population programmes to stress education, training and higher status for women as means to slow population growth. As the case of South Korea clearly demonstrates, these conditions are most likely to be realized where income distribution is relatively even. In Korea, urban incomes had been levelled by war and rural incomes by a far-reaching land reform programme. The land reform, which took place as part of the country's liberation from Japanese rule, had led to the emergence of much small-scale rural enterprise. The period from the late 1960s to the end of the 1980s was a period of high real per capita growth, averaging more than 6% per annum. But the fertility decline in Korea between 1960 and 1974 was one of the fastest recorded to that time: the crude birth rate fell in those years from 41 to 24. Birth rates started to decline at about the same time in all regions and among all income groups. In the first five years, too, the availability of artificial contraception was relatively unimportant in this decline, although it became more so subsequently. By 1973 artificial contraception was being used widely, with the most rapid increase among less well educated women; rates of abortion, previously high, though then illegal, began to level off. In parallel, as prosperity spread, children gradually came to be viewed less as an economic resource and the ideal family size fell. Universal education meant a much heavier investment in each individual child. Last but not least, greater employment opportunities, especially for women, resulted in a later average age for marriage, thus further tending to limit family size (Repetto, 1979, pp. 69–116).

Apart from Hong Kong and Singapore, the only other country in which a comparable decline in fertility took place in this same period was Taiwan. In 1949, Taiwan was poorer than the mainland of China, although it did have a small established industrial base dating from its period under Japanese rule in the 1930s. Between 1949 and 1955, land reform limited the size of individual holdings. Although generous compensation was given to the larger landowners, who used the funds to start their own businesses, the distribution of wealth was more even than in most Southern states. High interest rates encouraged high domestic savings and foreign borrowing was unnecessary, but because of its strategic location Taiwan did get an enormous amount of US aid. State-led economic development placed the highest priority on infrastructural development and the expansion of education (Harris, 1986).

Studies of the impact of development programmes on fertility in Sri Lanka and Thailand showed that they had a direct effect and that it was negative, as expected. However, the impact of individual programmes in this way was low compared with the overall impact of development (Stoeckel and Jain, 1986, pp. 8–9).

Despite this evidence, population remains a contentious issue. Not all countries are seeking to reduce growing populations: some still see the status and development of their countries as dependent on precisely the opposite process. Argentina has never really moved from a policy of 'peopling the Pampas'. It is an enormous, resource-rich country, which could sustain a much bigger population. Likewise, Bolivia, Iraq, North Korea, the Congo, Laos and (paradoxically) Singapore are all countries with governments that are seeking to increase population size. Again, the nature of environmental problems confronts an outmoded state-centric thinking that quite frankly does not address it.

A global approach is necessary. The fact that a relatively egalitarian economic structure affects national rates of population growth has important implications for the future of the world's population as a whole:

> The world fertility rate is affected by the international distribution of income in the same way that national fertility rates are affected by income distributions within countries. The same reasoning and kinds of evidence used in interpreting national fertility rates are applicable to the world rate, and lead to an analogous conclusion: a more equal distribution of income among countries would be conducive to a lower global birth rate and a lower rate of world population growth.
>
> (Repetto, 1979, p. 156).

HEALTH AND MORTALITY

At present, one-quarter of the earth's population is not getting enough food. Malnutrition begins before birth, as is shown by declining birth weights in many individual cases and over some areas. In Africa south of the Sahara, deaths of under-fives contribute between 50 and 80% to the total mortality of the population, compared with 3% in Europe, reflecting their malnourishment, lack of access to clean water and other consequences of poverty.

Infant mortality rate (IMR), defined as the number of children per 1000 who die in the first year of life, varies strikingly between the major regions of the world. The world average is 81 per thousand. This compares with, on the one hand, Europe, where the average is 16, and, on the other, Africa, where it is 114. These figures conceal striking variations within regions. Likewise, IMRs are always higher in rural areas, which are less likely to have the same levels of access to medical services, female education, potable water and proper sanitation, or the incomes necessary to achieve adequate levels of nutrition. Malnutrition has a serious effect on the unborn child, and in acute cases or if continued throughout development can lead to irreversible impairment. Surveys in East Africa and South Asia show that children under five are 'moderately' malnourished in some 15–30% of cases, and the same percentage of children have low birth weights, indicating probable malnourishment during pregnancy. At the same time, poor hygiene and unsafe storage conditions make food poisoning a serious hazard for both children and adults.

Malnutrition, aggravated by infectious diseases spread by poor sanitation and polluted water supplies, causes the bulk of Southern mortality, especially in children under five. Of the 15 million unnecessary infant deaths each year, 4 million are from one or more of six cheaply immunizable diseases and a further 5 million result from diarrhoea preventable by oral rehydration therapy, the salts for which cost next to nothing. The cost of just three weeks of what the world's governments spend on arms would pay for primary health care for all Southern children, including ensuring access to safe water and immunization against the six most common infectious diseases.

In the past, lack of agreement on the way diseases such as cholera and yellow fever spread delayed a concerted response by the international community, Richard N. Cooper estimates, by three-quarters of a century (Putnam and Bayne, 1987, p. 9). The legacy of this is still with us, since the delay meant that only in the case of smallpox has an international campaign so far actually succeeded in eliminating the threat altogether. Inadequate funding means that the disfiguring disease leprosy, which is hard to transmit and easily cured, still damages human lives. There has been an alarming revival, too, of one of the most potent killers, tuberculosis, in the USA, where health programmes are inadequately funded under the influence of 'New Right' economics.

In addition, women in Southern countries are more likely to die of the complications of childbirth or of the consequences of a botched abortion. And when effective means of contraception are lacking, the traditional remedy for an unwanted pregnancy is either abortion or infanticide. In Bangladesh, 750,000 sought to terminate pregnancies in 1978. According to the Center for Disease Control, of these at least 7500 died and thousands of others were left sterile or seriously injured (Moffett, 1994, p. 199).

FAMINE

There can be no question that famine constitutes the ultimate environmental disaster for those vulnerable to it. The cruellest way to limit population is to starve people to death. The failure of crops is at least one of the causes of mass starvation. Dando (1980), however, regards it as the prime cause in only one-third of the historical instances he analyses. If famine is an event, it is an extreme case of a larger process, and poverty and war (to name only two) are as much recurrent factors in the story of famine as crop failure. Analysis of 800 famines spanning 6000 years leads Dando (1980, p. 87) to identify five main types:

1. *Physical famines*, where the environment was naturally hostile, but techniques were developed which enabled human beings to overcome natural hazards in all but occasional cases.
2. *Transportation famines*, where highly urbanized regions depended on a well organized supply system, which from time to time went wrong.
3. *Cultural famines* in regions with a food surplus where archaic social practices or cultural prejudice ensured that it was not distributed.
4. *Political famines*, where regional politics determine the production, distribution and availability of food.
5. *Overpopulation famines*, where overpopulated, marginal agricultural regions subsist only just above starvation levels.

The best known physical cause leading to famine is drought, which causes crops to fail and farm animals to go short of food. However, flood, which interrupts supply, can also be followed by famine if relief is not available or is delayed.

In modern times theorists have been at pains to argue that food shortage, as such, is neither a necessary nor a sufficient cause of famine (Devereux, 1993, p. 11). Devereux argues that dictionary definitions of famine are based on 'Western' (i.e. Northern) perceptions of famine as a crisis of mass starvation containing three elements: food shortage, severe hunger and excess mortality. Famines, he argues, can occur without any one of these conditions being fulfilled. Moreover, famine victims do not see famine as an event at all. 'They see famine as (1) a problem of destitution, not just starvation, and (2) a continuation of normal processes, not a unique event' (Devereux, 1993, p. 181).

As noted above, the classic explanation of famine was that of T. R. Malthus, who argued in 1798 that it was the necessary and inevitable consequence of population outstripping the productive capacity of agriculture. Excess population, in his opinion, was the cause of famine, and today neomalthusians argue that famine occurs when growth in population outstrips the carrying capacity of land. In modern times this theory has been unfashionable, especially in the optimistic years of the green revolution, since when a surprising number of apparently sensible people have seriously argued that agricultural production will always outstrip population increase.

Neoliberals are 'counter-Malthusian' in their optimistic belief in the ability of the market to solve all problems. For them the problem is one of market distortions, of insufficient or irregular food supply; the remedy is to enable farmers to grow crops and make the necessary supplies available. Ester Boserup is a 'counter-Malthusian' in a different sense, in that she argues that increased population brings not only increased demand for food but also increased productive capacity, resulting from the possibility of improving infrastructure and so productivity. For her, population is a resource and crowded nations are productive nations, and this is a view widely held in Brazil and other Latin American states.

Amartya Sen rejected both population increase and crop failure as explanations, arguing that trying to explain famine in terms of 'food availability decline' was at best incomplete and at worst wrong. For Sen there is enough food in the world (at present) to feed everyone; the problem is that so many of them lack access to it, owing to power structures. 'Starvation is the characteristic of some people not *having* enough food to eat. It is not the characteristic of there *being* not enough food to eat. While the latter can be a cause of the former, it is but one of many *possible* causes' (Sen, 1981, p. 1). Sen's distinction between the availability of food and the individual's 'entitlement' to it is crucial. Entitlement has four sources. Trade-based entitlement gives access to food through commodity exchanges. Production-based entitlement gives a right to own what one has produced with one's own resources. Own-labour entitlement derives from the sale of one's labour power to others. Inheritance and transfer entitlement is the right to own what has been freely given by others.

> A person is reduced to starvation if some change either in his endowment (e.g. alienation of land, or loss of labour power due to ill health), or in his exchange entitlement mapping (e.g. fall in wages, rise in prices, loss of employment, drop in the price of goods he buys and sells), makes it no longer possible to acquire any commodity bundle with enough food.
>
> (Sen, quoted in Devereux, 1993, pp. 67–8).

One of the problems of defining famine is that starvation is at least partly a question of choice, and in famine situations individuals often choose to go hungry rather than to sell vital assets. Moreover, in India, for example, hunger is endemic, so that Sen has calculated that some 3.3 million people die in that country from starvation each year – as many as died in the Great Bengal Famine of 1943. In such circumstances it may be difficult to determine just when the onset of a famine can be expected. However, a modern government in control of its territory expects to watch for the signs of crisis in either production or distribution that could trigger a famine, and to seek to avert it by making supplies of food available in the affected region. Following the famine in Bangladesh in 1974, the government of Bangladesh was able successfully to avert potential food crises in 1979 and 1984 (Devereux, 1993, pp. 138, 140).

FOOD AND THE POLITICS OF FOOD

The Bangladesh famine of 1974 was to date the latest 'crisis of mass starvation' in Asia. The case raises another issue: the dependence of modern states on imported food. Since time immemorial, failure of crops in one region has been followed where possible by attempts to buy food from areas not affected. In 1974, however, the loss of some 12% of the Bangladeshi rice crop as a result of flooding came after the failure of six successive harvests in the previous three years, and was complicated by the aftermath of the war of independence. Speculators seized on the opportunity to profit by the crisis, driving up prices beyond the reach of the poor, while the government, by targeting assistance on urban areas and politically important groups like the army and the police, made matters worse. Lastly, at a time when famine in Africa was already commanding world attention and driving up world prices, the US administration of President Richard Nixon not only refused to send aid to Bangladesh until minor political concessions had been made, but in the process even withheld its normal allocation of food to Bangladesh under the PL 480 food surplus programme (Devereux, 1993, pp. 174–6).

On the other hand, huge quantities of food were sent by the USA to Guatemala after the earthquake of 1976. Since the earthquake had not affected the harvest, the effect of this was to drive down prices and to create a new stratum of dishonest traders who profited from the situation.

Sen is right that there is enough food in the world, at present, to feed everyone. However, the argument that the general problem of hunger is one of lack of entitlement is less clear-cut. Although in a sense the state-centric interpretation of notions of 'entitlement' constitutes a large part of the problem, the situation with regard to food is very special: it is the USA, Canada and Argentina that produce the gigantic grain surpluses, but the rest of the world depends largely on indigenous resources for its basic staples. Indians eat mostly rice grown in India, Thais rice grown in Thailand. Vietnam used to produce surplus rice for export; now even Thailand is running short. But although many Indians do not have enough rice, large packs of high-quality, Indian-grown basmati rice can be found in any Northern supermarket. Mexico now imports maize from the USA. When a country such as Mexico no longer produces enough of its basic staple to feed its own population, alarm bells should ring. With twice the population today that it had in the 1950s, India could no longer look to the USA for relief and be sure that it would come.

The drive to find and exploit crops for export has resulted in a worldwide move from the growth of foodstuffs for subsistence to the production of cash crops for export. Such cash crops now occupy more than a quarter of the cultivable land in the developing countries. This shift has serious consequences. First, it increases people's vulnerability to famine because personal reserves of food no longer exist. Second, for the individual family, it introduces a new kind of vulnerability: dependency on macroeconomic changes. The periodic fall in commodity prices, which previously would have affected only a few, now acts to reduce the country's foreign exchange earnings and hence its capacity to buy staple foodstuffs which are no longer locally produced in sufficient quantities. These pressures on the people of developing areas are also pressures on the often fragile environments they inhabit.

Africa south of the Sahara was a food exporter until 1960.

According to Shiva (1989), the region was still feeding itself as late as 1970. But by 1984, 140 million out of 531 million Africans were being fed with grain from abroad. The main reason for this is that export-oriented cash crop production has replaced subsistence farming.

On the other hand, no matter how wealthy you are, there is quite a modest limit on the amount you can eat. The wealthy do not eat so much that it stops there being enough to go round. They eat different foods that are rarer or more expensive, and frequently more wasteful in ecological terms. Such foods are often more expensive because they have to be hunted: wild salmon, larks' tongues and caviar all come from finite populations of wild creatures. As long as this is the case, there is a delicate ecological balance to be maintained between having enough of the product to satisfy demand and doing irreparable damage to the population.

Northern views of famine in the South tend to have an unspoken assumption that the cause lies in the South. Famine, like other natural disasters, is seen as a situation with which the local inhabitants cannot cope. However, the evidence is quite unequivocal that this view, explicit or implicit, is wrong. Most of what is done to help people in the aftermath of a natural disaster is done by the people themselves. The 'myth of helplessness' has now long since been rejected by Northern aid agencies. However, in doing so they have not found it easy to decide how best to target their efforts.

> It is possible to make three broad generalizations about disaster aid from the North. First, it is highly variable and follows no logic of need and cost-effectiveness. Second, it tends to follow the pattern of the donor country's development and military aid; i.e. relief aid tends to go to allies and those already getting aid. Third, a lot of 'relief' is merely the export of surplus commodities.
>
> (Wijkman and Timberlake, 1984, p. 112).

The most common hazards, drought and flood, attract the least support from the North, most of the aid going to the victims of civil war and earthquake.

Food aid can save lives provided that it is properly targeted and does not continue for so long that the people become dependent on it. Unfortunately, there is also a great deal of evidence that in disaster situations the food does not always go to those most affected, and that if it does, it can drive down the price of locally produced grain, giving farmers a strong disincentive to plant for the coming season. In emergencies, of course, people will eat anything. However well meaning relief efforts are, though, they often run into the problem that the local inhabitants simply have no experience in using strange food; they do not know how to cook it or they lack even the basic resources (fuel, clean water) to do so.

THE POSSIBILITY OF EMPOWERMENT

Rapid population growth does harm the life chances of the poorest members of society; it reflects their fear of economic insecurity but at the same time deprives them of access to land, food and jobs. It is women who suffer disproportionately from the effects of rapid population growth. Availability of family planning cannot of itself slow population growth, except in a social context in which the decisions are made by men and women (but more particularly by women) themselves (Lappé and Schurman, 1989, p. 55).

The capacity to take such decisions, to control one's own life, is the vital ingredient in population control and sustainable use of local environments more generally. To start with, capital must be available to people who in the past would not be regarded as qualifying for it. The Grameen Bank, founded by Muhammad Yunus in 1982, and BRAC, founded by F. H. Abed, both in Bangladesh, provide microcredit to villages for small enterprises. Regular attendance at a weekly empowerment group to encourage self-education and thrift is a condition of getting a loan, and the repayment rate is over 97% (Moffett, 1994, p. 204). But there is a long way to go before microcredit is available to all who could make use of it.

The next need is a source of friendly and helpful advice. Here there is an important role for NGOs at all levels. NGOs are able to operate internationally, offering aid and advice without the problems of political interference and suspicion that surround official government action. Obviously there is a need for NGOs not to confront national governments where this would be counter-productive, and hence in places such as Burma no useful work can be done. However, governments may be more prepared to accept their own nationals as NGOs. At the local level especially they can give practical help in empowerment of individuals and small local groups, creating the structures of civil society that enable people to live better. Given the caution of all governments in dealing with the issue of reproductive rights, there is a special role for NGOs in this field.

CONCLUSION

Urbanization and industrialization go together. From the beginning of the industrial revolution, employers have found it convenient to be able to draw on large concentrations of cheap labour. However, at the end of the millennium there are signs that this, at least, may be about to change. As European firms depend on data processing in India, where skilled labour costs are lower, they have demonstrated that it is no longer necessary in all cases to live near one's place of employment.

Certainly, the city points up the lesson that individual poverty is a major problem for the environment. Poor living conditions lead to organic pollution of water and land alike, and the rapidly growing cities of the South show clearly the link between poverty and disease. But the cities of the South are also victims of some of the worst atmospheric pollution anywhere in the planet. A compensating factor, perhaps, is the fact that poverty in the city is relatively easy to see, and it is the squeaky wheel that gets the grease. In any case, governments like to spend money in the cities because of the quick pay-back in terms of votes and political support. Urbanization in the USA in the late nineteenth century gave a great impetus to the extension of democracy, and in time the same phenomenon may do the same for the South.

By contrast, rural poverty is widespread and commands much less political attention. Indeed, Southern politicians benefit electorally from controlling the flow of benefits to the rural masses. The same factors that lead to the growth of cities lead to the deterioration of the countryside, exacerbated by the urban bias of ruling elites. The countryside is not a favourable place for the growth of democracy, and its environmental problems, being long term and slow to change, do not command public attention unless disaster supervenes.

In both the cities and the countryside, poverty persists throughout the South, and in turn the deterioration of the environment, of which it is the principal cause, increases poverty. Malnutrition, disease and death are the most visible consequences, but their persistence is made worse by the long-term effects that they cause. Occasionally disasters call the attention of the North, but the famine and pestilence that accompany them are also the product of long-term causes as well as short-term crises, and of political causes, such as the maldistribution of wealth and the powerlessness of the poor, as well as physical changes in the environment. Finally, the persistence of poverty imperils the survival of the whole human species, owing to the steady decline in the 'carrying capacity' of the land resulting from land degradation, salination and desertification.

CHAPTER 8

THE INTERNATIONAL POLITICS OF THE ENVIRONMENT

RELATIONSHIP OF NATIONAL AND INTERNATIONAL STRUCTURES: WHO GOVERNS?

The dilemma that the environment is international, but power remains with nation states, has been recognized by many writers. 'How can international institutions, which necessarily respect the principle of state sovereignty, contribute to the solution of difficult global problems? What are the sources of effectiveness for institutions which lack enforcement power?' (Haas *et al.*, 1993). Given the assumptions of the international system, it is not surprising that there was no real response at international level to transboundary environmental problems before the United Nations Conference on the Human Environment (UNCHE) at Stockholm in 1972.

The Stockholm Conference originated from the North's concern about pollution. At an early stage it was already clear that the South was suspicious about Northern motives. At the meeting in 1971 of the Preparatory Committee (PrepCom) at Founex, Switzerland, the Northern delegations agreed that lack of development was in itself an agent of pollution, and had accepted that the Southern states had legitimate concerns in seeking to emulate the pattern set by the North. In the *Stockholm Declaration on the Human Environment*, accordingly, the 26 principles set out included specific recognition that, given 'rational planning' and/or 'integrated development', development could be achieved without harm to the environment. However, neither the Declaration nor the accompanying *Action Plan for the Human Environment* was specific about what would constitute either of these two desirable goals.

The more important consequence of the Stockholm Conference, therefore, was the establishment of the United Nations Environmental Programme (UNEP), the first UN agency to have its headquarters in a Southern state, Kenya. In its first 25 years of existence, UNEP has some important achievements to its credit, including raising environmental awareness and helping to foster national environmental protection, even if mainly in the AICs.

The *World Conservation Strategy* (WCS), published by the IUCN in 1980 with the support and assistance of UNEP, was the first major statement to employ the term 'sustainable development' and to try to clothe that concept in specifics. Its closing section emphasizes that development and conservation operate in the same global context and have to overcome the same problems.

> It identifies the main agents of habitat destruction as poverty, population pressure, social inequity and terms of trade that work against the interests of poorer countries. It gives a check list of priority requirements, national actions and international actions, and calls for a new international development strategy aimed at redressing inequities, achieving a more dynamic and stable world economy (to allow all countries to participate more fully and more equitably), stimulating 'accelerating' economic growth and countering the worst impacts of poverty.
>
> (Reid, 1995, p. 40).

The notion of sustainable development grew out of 1970s ideas of development, as moderated by the experience of the oil crisis of 1973. This also led to the call by various Southern states for a new international economic order and the adoption by the UN General Assembly in 1974 of the *Charter of Economic Rights and Duties* (United Nations, 1974). Proposed by Mexico, this marked the high water mark of 1970s developmentalism. Sovereignty, autonomy and non-intervention were the themes of the Charter. Sustainability was not mentioned. But Southern states, too, were hit by the oil crisis, and in 1977 the General Assembly of the United Nations set up the Independent Commission on International Development Issues (ICIDI), with the former West German Chancellor, Willy Brandt, in the chair.

The Brandt Commission's first report, *North–South: a Programme for Survival* (Brandt, 1980), accepted that development based on simple considerations of economic growth had failed to secure 'human dignity, security, justice and equity' for much of the world. However, it insisted that without economic growth and industrialization the standard of living in the South would not improve. While accepting that bad government decisions in the South had contributed to this situation, it placed most of the blame on world recession, high interest rates, falling prices for primary products and Northern protectionism. Its remedy was to transfer more funds to the South (it was here that the target of 1% of GNP was first proposed), while at the same time reducing the comparative advantages enjoyed by the North. Not surprisingly, in the climate of the 1980s, it failed to win support in the Northern states, which alone could provide the necessary funds. After its second report in 1983 the Brandt Commission was dissolved, and in the same year the General Assembly set up a new World Commission on Environment and Development (WCED), chaired by the Prime Minister of Norway, Gro Harlem Brundtland. The numerical majority of the Southern states in the UN was reflected in the choice of the other 22 members.

The new Commission recognized that the twin concepts of environment and development had to be addressed together. It found the solution to the apparent contradiction between them in the concept of 'sustainable development'. The Brundtland Report, *Our Common Future* (WCED, 1987), defined sustainable development as: 'development that meets the needs of the present without compromising the ability of future generations to meet their own needs'. Sustainable development does not use non-renewable resources faster than substitutes can be found, or renewable resources faster than they can be replaced; nor does it emit pollutants faster than natural processes can render them harmless.

However, sustainable development, remains about development not just sustainability. Underdevelopment was not a term invented by Harry Truman, and he did not use it to establish a new colonialism — although Gustavo Esteva's argument that this is what happened cannot be ignored (Esteva, 1992). Ever since human beings have been on the planet they have been consuming and using natural resources, and the idea that the earth's resources are limited is a very hard one for most people to believe. Hence, no proposals that deny Southern countries the chance to develop can possibly work, while the AICs will not easily yield anything of their present standard of living.

At the moment, unbalanced development is a major cause of the destruction of the environment. The poor can hardly be blamed for exploiting their environment in a way that destroys its long-term potential when the rich, including the TNCs, are doing the same thing on a much more massive and destructive scale. Deteriorating terms of trade and the burden of debt on the Southern states increase pressure for exploitation by making short-term returns the most urgent consideration. The implication is that, in any plan for sustainable development, the future development needs of the South must be allowed for. Development cannot be sacrificed for environment any more than the other way round. The North–South gap cannot be perpetuated if the world is to remain stable. Therefore, sustainable development must meet the economic, social and environmental needs of all the world's peoples.

Brundtland's call for sustainable development is now an article of faith in the First World, but too often it has been seen as open to a much more limited interpretation, that of growth as usual but slower. The Brundtland Report itself is a compromise. It is weak on the subject of population growth and does not really suggest a solution to the problem of rich world resource consumption. At present, the depreciation of natural resources is not taken into account in calculating GNP and other indicators. Economists have already been able to demonstrate 'the physical dependency of economic activity on the sustainability of crucial natural-resource systems and ecological functions, and to indicate the economic costs, or trade-offs, resulting from the failure to preserve sustainability and environmental quality' (Barbier and Markandya, 1989, p. xiv). If new measures were to be devised, it would be easier to see where sustainability was attainable and where not. Academics therefore have a role to play in getting people to see the world differently.

THE UN AND THE ENVIRONMENT

The role of the UN is questionable. Since the 1980s, a continual preoccupation has been whether it will continue to get funding if it does not 'reform' in the way that its largest financial contributor, the USA, requires. But the UN in any case falls far short of being a world government. There has been disappointing progress in global approaches to any of the other major issues the UN has sought to address: the persistence of poverty, action on health, the status of women. The UN is really the only organization that can organize concerted action to be taken by the majority of states, but it lacks any power to compel action on the environment even where it has been agreed.

It was the Stockholm Conference on the Human Environment, with its slogan 'only one earth', the first of many large international conferences, that added the environment as a huge global policy problem to the UN agenda (Soroos, in Vig and Kraft, 1994, p. 299), and set the scene for the signing of a series of key international treaties. Previously, single environmental issues had been dealt with by different UN agencies. Just as Stockholm led to the creation of UNEP, so its concurrent Environmental Forum brought (accredited) NGOs formally into the process of environmental policy-making for the first time, and set the pattern that Global Forum would follow at Rio.

UNCED was a consequence of the deepening environmental crisis, which was increasingly subject to investigation and thus more understood as a result of scientific progress. It was this understanding that led to the Brundtland Commission publishing *Our Common Future*, which provided the backdrop to the preliminary meetings that were to culminate in the 1992 UN Conference on Environment and Development in Rio de Janeiro: UNCED, also known as ECO 92 or the 'Earth Summit'. It was the largest gathering of world leaders ever held, bringing together representatives of 179 states, including 118 heads of state in person, and it put ten related issues squarely on to the global environmental agenda.

UNCED recognized the interconnectedness of the environmental and developmental crises. Nevertheless, the stress placed upon these two aspects of the Conference by participants varied. Developing countries emphasized their crisis of development and international economic inequality as the main source of the environmental crisis. They expressed their fears of environmental imperialism. The Northern states were concerned about environmental issues but generally did not want to meet the costs in financial or political terms of either extending existing international regimes or negotiating effective new ones (Spector *et al.*, 1994).

Rio was oversold: its slogan was 'our last chance to save the earth'. The image of Cristo Redentor (the statue of Christ the Redeemer which stands on the Corcovado, high above Rio) was evoked as a mark of the redemptive qualities of the meeting. It was hoped that putting priorities into one document for the twenty-first century, *Agenda 21* (United Nations, 1992), would lead to optimism and the will to do something; thus finance would be forthcoming. This hope has so far proved illusory. In fact, the situation has got considerably worse. The hope of generating the necessary funds had failed before the Conference even met. The EU made the only substantial financial commitment, an extra US$3 billion, but it has not delivered. The OECD has spent less than before. OECD aid, US$60 billion a year before 1992, is now US$50 billion a year, and private aid is now twice official aid.

SOUTH AND NORTH

Structural constraints enable the North to control international debate by controlling the agenda and manipulating the processes within which it is discussed. Is the ecocratic discourse, the language of environmental politics, itself Northern dominated? This question is central to Southern challenges from within environmentalism, such as that of Gudynas (in Sachs, 1993), and from without environmentalism from the left, such as Durkin's *Against Nature* series (Channel 4 TV, 1997; see also The *Guardian*, 2 April 1998). The South charges the North with 'ecoimperialism', arguing that Northern colonial attitudes are nowadays expressed in ecotourist visits to human zoos and (perhaps worse still) to wildlife reserves that exclude those people who previously made their livings there. As Gudynas notes, there is a tendency to reduce and so to distort the debate. It easily degenerates into an international tussle over the local environment of the pretty bits of the world or the parts perceived as of global importance, most notably Amazonia. Other environments are forgotten and neglected in the process of prioritizing, and some parts of the world become associated with only a single environmental issue.

These different interpretations are reflected in the two different uses of the term 'commons'. The basic idea is that there are certain parts of the land or sea that are not the private property of any person, company or state. However:

1. Pro-Southern perspectives tend to use the term as those areas and resources upon which the poor who own nothing rely. Such is its use by Hildyard (in Sachs, 1993).
2. The term has also been used by environmentalists to express the common inheritance stressed by the North at Rio and rejected by many Southern leaders, though not by all (*The Ecologist*, 1993). Southern interests have argued that the use of the term 'global commons' is effectively the moral hijacking of their resources, another example of ecoimperialism. The Northern use of the expression 'global commons' has had to confront Southern national sovereignty arguments, such as Dr Mahathir's assertion that if the North wants to benefit from Malaysian forests it must pay for them.

It might have been easier for the South to swallow Northern arguments about a common heritage and a common future if the North had not seemed quite so implacably opposed to reducing its own consumption. The importance of this issue was exemplified by the domestic policies adopted before and after the US presidential election in November 1992. Bush's contribution to UNCED earlier that year was to state at the outset that US lifestyles were not negotiable; he had earlier described US environmentalists as extremists seeking to put ordinary Americans out of work.

Perhaps the real dichotomy in the use of the term 'commons' is illustrated by the reluctance of the USA to sign an accord (biodiversity) that might have constrained American-based TNCs from continuing to enclose (through patents) precisely those 'commons' on which the world's poor depend. At UNCED, the global commons (and in particular rainforests) were amongst the main problems identified by the North. The others were climate change, the ozone layer and population. On the other hand, the countries of the South sought to emphasize a different agenda: poverty, hunger and desertification.

Three new Conventions were agreed as a result of the Rio Summit – on climate, biodiversity and deserts – along with a new set of guidelines for forests. The Framework Convention on Climate Change was signed by 155 countries (Patterson, 1993). However, its commitments were inadequate, especially those made (or not made) by developed nations to limit greenhouse gas emissions. The IPCC estimated that 60% reductions were necessary to stabilize carbon dioxide concentrations at their 1992 level. The word 'aim', however, was chosen, as US negotiators freely admit, so as *not* to be a commitment, and the target date of the year 2000 was also dropped from the sentence at the Bush Administration's request. In the event, the wording was much weaker than the Alliance of Small Island States wanted or the EU was then prepared to accept.

North–South divisions focused on transfers of technology and funds that were needed if developing nations were to meet any of the Convention agreements. The dependency on those transfers was included in the Convention, but no commitment was made to make the transfers. The North agreed to provide financial assistance in compiling national inventories of 'sources' of carbon dioxide and the 'sinks' that remove it from the atmosphere (see Chapter 2), but did not make any arrangement to meet the costs of actually doing so. The problem is that such agreements are consensus documents and therefore tend to be minimal in terms of commitment. What was needed was consensus not just between different perceived interests within developed countries, but between the rich industrialized world and the developing states, not to mention between the hugely varied interests within those developing states. The consensus was very limited, and current levels of commitments will not slow global warming.

Despite its weaknesses, however, the Framework Convention provided evidence of what is needed (and known to be needed) to slow climate change. It provided a framework for future discussions and an institutional structure for them, leading up to the Kyoto Conference in 1997. Since this in turn failed to make any real progress, the entire problem was again remitted to Buenos Aires in November 1998 and this in turn agreed to put off any action until long after any of those present was likely to be in the office. But very small climate change would produce changes that are generally seen as likely to impact particularly heavily on the developing world. For example, drought risk is not equally spread, and very small changes in drought risk will produce big impacts on agriculture, leading to significant agricultural losses in Brazil, Peru, the Sahel, South-East Asia, the Central Asian Republics and China. Sea level rises could potentially cause massive people movements. Relatively small rises in mean sea level could displace 30 million people in Bangladesh, for example.

There has been some success in agreeing the drylands agenda, but there is no new funding to ensure that it becomes reality. Carbon dioxide emissions are higher than when the Convention was agreed. One of the rare exceptions to this among the AICs is the UK, where carbon dioxide emissions have been cut during the 1990s. This, sadly, was not because of enlightened environmental policies, but the decline of the British coal industry, combined with recession. The USA, Australia, Canada and Japan are seeking to evade any reduction in their greenhouse gas emissions by buying quotas from the republics of the former USSR (including Russia), where economic collapse has left them well below 1992 levels. If they are successful, this will mean that there will be no net reduction in world emissions. Meanwhile, species and habitats are disappearing more quickly than they were in 1992, and so are forests. UNEP should be monitoring what is happening, but it is all but moribund.

Worse still, the completion of the Uruguay Round of the General Agreement on Tariffs and Trade (GATT) resulted in the creation of the new World Trade Organization (WTO), and the limited perspective already shown by it has gone far to destroy any trust that Rio might have engendered in the South. The stress on markets has intensified competition and put further pressure on the environment. It is not just globalization working against the Rio aspirations, but the nature of the fragmentation of international and national agendas. The UN is increasingly seen as unimportant by the wealthy developed nations. It is in internal turmoil and confused about its role. In key respects it has been successfully by-passed by the G8, the club of rich nations. Even at the national level, conflicting interests abound. Governmental good intentions are difficult to achieve across all national departments: for example, transport, trade, defence, education, health and finance all have their own goals and objectives (Allison, 1971).

In June 1997, the representatives of national governments reconvened at the UN General Assembly in New York for a review of progress since UNCED. This United Nations General Assembly Special Session, or 'Rio plus 5', lacked the sense of purpose of the original. Only a few heads of state or of government attended, and it was hard not to feel disillusioned. But Rio was not a complete waste of time and resources. Rio has in some respects at least helped to change public perceptions in the developed world. 'Green' investment, for example, now amounts to some US$600 billion worldwide, and the potential is hardly being tapped. As part of this development, the impact of Rio on business has been the increasing awareness of the negative images among consumers which environmental protests can inspire. In August 1997, for example, in response to interference with oil rigs in the North Atlantic, BP issued writs against Greenpeace and had the organization's UK bank accounts frozen. However, all BP got for its trouble and expense was a good deal of bad publicity. The company was forced into a very undignified retreat. Since then, in 1998, Shell has withdrawn publicly from the misleadingly named World Climate Coalition, the objective of which is to fight off any attempt to reduce consumption of fossil fuels. Clearly it did not do so without careful consideration.

GLOBAL PROBLEMS/LOCAL ACTION: WHERE DOES POWER LIE IN ENVIRONMENTAL MATTERS?

One of the main failures of Rio was the lack of concrete proposals for addressing the problems identified. Of course, this was an obvious consequence of state-centric structures seeking compromises between very different actors with their own agendas and constituencies. Brundtland's adage that we should 'think globally and act locally' was, however recognized in the special role accorded to women and indigenous people in Agenda 21, and perhaps the most impressive implementation of Agenda 21 ideas has been at the local level. It is possibly at the local level that the impact of Rio is most felt. Local authority Agenda 21 schemes can now be found not only in Sweden, where environmental awareness is most developed, but also the UK, where two-thirds of councils have signed up to them. The level at which environmental solutions must be sought is highly contentious. Is there still a role for the nation-state, or is it, as Daniel Bell suggests, 'too small for the big problems of life, and too big for the small problems of life' (Bell, 1987)? The Prince of Wales talks of 'primary environmental care', meaning that in the long run simple and local solutions are most important.

Collectively, individuals as consumers of TNC products are important, and the local implementation of environmental protection remains vital, but what of the wider level? In considering how potent environmental issues are in power politics Lukes's three-dimensional model is illustrative.

Lukes (1986) identifies a first level of power, which can be measured by simply counting the outcomes that favour the interests involved, as in Dahl's study of New Haven, Connecticut (Dahl, 1961), and Christopher Hewitt's 'issue method' (Hewitt, 1974). Football provides a metaphor: this first dimension would simply involve counting the goals. Certainly, environmental groups are winning some issues locally, nationally and internationally.

At the second level, Lukes was concerned with the capacity to get issues on to the political agenda or keep them off it: in footballing terms, which games get played (the draw). It is no coincidence that this level of power is often exemplified by Mathew Crenson's 1971 study of how US Steel and other large companies managed to keep air pollution off the political agenda until well into the 1960s in some US cities. It is no longer possible to keep environmental issues off the agenda in such a crude way. This fact has led to a new approach to the control of the agenda by TNCs, as illustrated by Monsanto's opening up of a debate in its advertisements for its genetically engineered products.

The environment is certainly now on the agenda, but what about the third dimension of power in the Lukes model: in footballing terms, the slope of the field? The global economic system and the needs of the world's population are in conflict with each other, and neither necessarily harmonizes with environmental needs.

There is a obviously a tension between the natural and the social and the economic. It is precisely economic development and the social changes that have gone with it that have raised the costs to the environment in the present, and more especially in the very near future. The question of whether 'green growth' is possible is a highly contentious issue.

But environmental crises exist at different levels, from the global to the local. Such crises may involve this further tension. For example, regional economic development may confront local resistance to environmental damage. There are different pressures on the environment in different places: economic development and especially the high environmental costs of industrial processes in the North. This is compounded by an international economic system resting on separate and competing states and an absence of global consciousness. As we have seen, this form of resource consumption coexists alongside arguments concerning the pressure of population on natural resources, especially in the South.

The scale and diverse causes of environmental degradation are part of the problem, along with uncertainty and the invisibility of aspects of the crisis. These factors work against recognition of the problems and against action to solve them. Further, no solutions will 'work' in a political lifetime and 'national' interests coincide with global ones only in the long term, if at all.

There is a need for increased global consciousness, and this must rest on a recognition that all economic development, wherever it has occurred, has always required inputs from other areas. We need the resources of other states, some more so than others. For example, the USA is resource-rich and Japan resource-poor and heavily dependent on imports. But both shamelessly exploit the environment without recognizing the ecological dependencies that must come to constrain our impact on our world. Last, but not least, global action requires local implementation, and the question of what form this takes is conditioned very much by the specific conditions of individual Southern states.

POLITICAL AND SOCIAL STRUCTURES IN NEW STATES

Much of the period since 1960 has been dominated in international politics by the Cold War. Although almost all the new states that emerged during this time paid at least lip-service to the ideal of democracy, in practice most if not all were dominated to a greater or lesser extent by an oligarchy. The elite to which power was transferred at the moment of independence, or which seized power by force soon afterwards, was able to maintain itself by its control of recruitment. If it was challenged successfully by opposition pressure, the new government was distrusted by both the superpowers for different reasons: the USA, because it feared the loss of political support, and the USSR because it was uncertain about the real political intentions of the regimes that professed adherence to socialism. But neither of the superpowers at that stage seriously questioned the overriding importance of development. The new governments therefore found themselves able to pursue developmentalist policies largely free of internal or external pressures to limit the impact of development on the environment.

This was even more likely to be the case where the armed forces seized power. A wave of military coups spread across Africa and Latin America in the 1960s, and by the end of the decade most of the states in both continents were under military rule. Military intervention in politics in Southern states results from both push (propensity/ disposition to intervene) and pull (stimulation/provocation) factors. Push factors for military intervention include the ambitions of individual officers, factional disaffection and institutional activity said or believed to be in the 'national interest'. Pull factors include the association of the armed forces with military victories, a general perception of a lack of cohesion, discipline or stability in society and a specific perception by the armed forces of threats to the military institution or to the officer class, or to the dignity or security of the nation. Both may be needed to trigger an actual intervention, as, for example, when the breakdown of legitimate civilian government is accompanied by changes in the military institution. However, the consequences were very similar: weak governments, lacking legiti-macy, which were very sure of their own abilities to rule, and had every intention of continuing to rule in some guise or another.

The economic failure of a civilian government affects the armed forces both directly, by leaving less money for them, and indirectly, by alienating popular support for the government (Nordlinger, 1977). Economic instability affects the social groups of which officers are members, rather than the military institution itself, whose budget is generally protected. However, it may be important in encouraging them to take action. Economic failure is seldom cited as a major reason for a military coup, and more often than not it has been given as a reason for the replacement of one military government by another (Wiking, 1983, p. 116). However, despite their evident limitations, military rulers are easily persuaded that they have a wider mission to bring about the development of their countries.

The first example of this phenomenon, then termed 'Nasserism', was seen in Egypt after 1954, and the Aswan High Dam and Lake Nasser both survive as a monument to a military leader's desire to shape the physical environment of his country. However, it was in Latin America in the 'developmentalist' era of the 1960s that a new phenomenon emerged, starting with Brazil in 1964, by which the armed forces seized power with the open intention of staying in power for an indefinite period, long enough to bring about the forced development of their societies. This phenomenon is often termed 'bureaucratic authoritarianism', a term originally invented by the Argentinian Guillermo O'Donnell (O'Donnell, 1988), whose views derive from Marxism and more particularly from dependency theory. The term bureaucratic authoritarianism is open to objection, in that it seems to underplay the central role of the armed forces. For the present purpose, however, the important thing is that by assuming the mantle of developmentalism, the armed forces were able to cloak in nationalist rhetoric their programme for the internal colonization of their countries.

INTERNAL COLONIZATION

The urbanization of the countryside and the expansion of towns is only a special case of the control of land and its capacity to generate wealth. The industrial revolution was preceded, and made possible, by an agricultural revolution which increased the productivity of a given amount of land dramatically. Control of political power was used to take productive land away from country-dwellers and to create large estates. In turn, the industrial revolution gave rise to transport systems which made possible the conquest of the countryside by town-dwellers. As this control was extended in the twentieth century, it resulted in the industrialization of agriculture (agribusiness). The impact on the environment of the 'internal colonization' of the countryside can be plainly seen in the replacement of traditional subsistence agriculture by large farms or plantations, managed for export and/or commercial sale of a single product or limited range of products.

Decolonization, paradoxically, accelerated this process. *The Ecologist* noted, under the heading 'From colonialism to colonialism', how Nehru and his successors had rejected the Gandhian dream of agrarian self-sufficiency. They set out to industrialize India by export-oriented growth, and other new states in the course of time sought to do the same:

A process of internal colonization, as devastating to the commons as anything that had gone before it, was thus set in motion. Using the slogans of 'nation-building' and 'development' to justify their actions, Third World governments have employed the full panoply of powers established under colonial rule to further dismantle the commons. Millions have lost their homelands – or the land they made their home – to make way for dams, industrial plants, mines, military security zones, waste dumps, plantations, tourist resorts, motorways, urban redevelopment and other schemes designed to transform the South into an appendage of the North.

(*The Ecologist*, 1993, p. 39).

Power, of course, was the key, and in this process TNCs, banks and lending agencies had at all times the willing collaboration of local commercial interests. Political elites, too, depended in a variety of ways on the establishment and/or maintenance of international trading patterns. In particular, the maintenance and, indeed, extension of plantation agriculture owes much of its vigour to the very special role of land as a badge of social distinction, an insurance and a hedge against inflation. Such landholding of course goes hand in hand with the growth of cash crops for export. The rulers of the South, therefore, are not often seeking any alteration in the terms of trade.

Resistance to this process was met by force. As weak civilian governments were displaced or supplanted by the armed forces, military governments set about conquering their own countries. Southern armies typically saw the economic situation confronting them as a military emergency, imperilling their ability to defend the state against its enemies. Their militarism – whether in Asia, Latin America or Africa – had a pride in their prowess which did not necessarily derive from recent combat, as, in many cases, for geographical or other extraneous reasons, the opportunity had not arisen. However, the fact of independence implied an important historic role for the forces. They had the role of guardians of the state thrust upon them, in their opinion, because they saw themselves as the ones who had given birth to it. So when the civilians failed to run the country in the way that they expected, they used the excuse of national emergency to step in and assume power.

In Nigeria, for example, the military takeover of 1966 owed much to the persistence of tribal consciousness in the army, among the northerners who felt excluded by the commercially active and politically dominant Igbo (Ibo). But it has proved much easier to get the armed forces into politics than to get them out again, and Nigeria, which has repeatedly succumbed to further military intervention, is also now rated one of the most corrupt governmental systems in the world.

The irony is that in the NICs, the authoritarianism of the government (except in the case of Singapore) was soon tempered by the widening distribution of wealth and the mobilization of the people at large in the service of economic development. But one thing that binds together authoritarianism of these different kinds is the impact on the environment. The environmental record of the NICs has been appalling. In particular, traffic congestion and industrial pollution have made the growing industrial cities very unhealthy places in which to live and work. In 1997, a bridge in the centre of Seoul, South Korea, collapsed during the morning rush hour, unable to bear the weight of the traffic that clogs the streets morning and night. However, it is also true that as countries gain surplus wealth so they become more environmentally conscious, and in this respect the NICs have been no different. An increasing number of home-grown environmental NGOs are now working in cities like Bangkok, and a growing middle class that can afford to support their efforts does so.

ARMS AND THE ENVIRONMENT

A special environmental problem is presented by the widespread distribution of arms throughout the South, and the use of them in armed conflict. Where either or both of the superpowers intervened to maintain the peace, in Korea, Vietnam and Somalia, new environmental problems were created and existing ones became much worse.

The collapse of communism ended the moral justification for much US intervention. The change was not immediate: the new public vocabulary included phrases such as 'ensuring international stability' or 'the worldwide crusade for democracy'. But the new US Administration of George Bush did not show much respect either for national sovereignty or for democracy. Its intervention in Panama in December 1989, claimed to be enhancing the cause of democracy, resulted in several days of fighting in Panama City, in which hundreds died and many buildings were damaged. It was carried out in breach of the charters of both the UN and the Organization of American States (OAS), and was censured by the OAS. Then, in 1991, the decision after a long period of delay to launch Operation Desert Storm on Iraq was justified as defending what Bush claimed was 'the legitimate government' of Kuwait. Although Saddam Hussein had few friends even among Arab states, and his own government had formally recognized that Kuwait was not the nineteenth province of Iraq, the Kuwaiti government was certainly not democratic and Bush did not explain on what other grounds it could have been considered legitimate. In fact, the decision to go to war had much less to do with defending democracy than with ensuring that the combined oil output of Iraq and Kuwait did not pass out of Western control, while the unilateral US decision to end the war after only 100 hours has had long-lasting consequences for the stability of other smaller oil-producing states in this key region and, of course, for the environment of the Gulf region.

The USA is now a 'lonely superpower', constrained by its own internal divisions to tread a much more cautious course. Other countries have different agendas and in groups could be powerful, but Russia is torn by internal dissension and the decision of German Chancellor Kohl to recognize Bosnia has left the European Union (EU) divided and apparently impotent. The end of the Cold War may well mean that in the future there will be no superpowers to restrain their former client states when local conflicts threaten to get out of hand. In a bipolar world, in a sense every conflict matters to the two camps, as victory for one is defeat for the other. In an era of multipolarity, most Southern conflicts will not matter to the AICs at all. These conflicts will, of course, damage local environments, as much by their interference with everyday life as by deliberate damage. It is only in the longer term that they will contribute to the degradation of the environment that we all share. If regional leaders do not intervene, no one may do so, and this may be even more undesirable.

Since 1989 there has been a sharp change away from bipolar confrontation, but the world is still littered with the debris of Cold War; literally so in the case of the conventional and nuclear weapons of the former Soviet Union. Week by week the news of the interception of smuggled uranium consignments compounds the uncertainty about the future intentions of middle-range states such as Iraq and Iran. The US post-Cold War strategy is to keep enough nuclear weapons to confront 'any possible adversary', and this continued nuclear hegemony is thought by some to be a means to prevent proliferation. However, the current non-proliferation regime rests on a crucial misconception: that a small group of nuclear states can retain nuclear weapons while denying them to others. The emergence of a Hindu nationalist coalition with a different agenda in India in 1997 has shown the fallacy of these assumptions (PPNN, 1998). More alarming for more localized environments have been the proliferation of chemical weapons and the deployment of landmines.

NON-GOVERNMENTAL ORGANIZATIONS

The campaign against landmines reflects the importance of NGOs, through their capacity to sway public opinion in national and hence international politics. A multiplicity of NGOs is characteristic of the democratic governments of AICs. The environmental movement began among NGOs, acting in the first instance as pressure groups on their own governments. Since then, they have assumed a wider international role, working across formal governmental boundaries and getting involved with international organizations on global issues requiring global solutions; in particular, with the UN and its proliferating special agencies (Weiss and Gordenker, 1996).

At Stockholm in 1972, 134 organizations were represented. Since then there has been an explosion of activity. The problem is that not all groups are of equal weight. It is clear that a small number of well funded transnational organizations enjoy special access to and influence with the UN. They act not only in public but in the continuing lobbying process that the international system requires. The main ones active at the global level are Greenpeace with 6 million members in 30 countries), the WWF, spawned by the IUCN in 1961 (now with 3 million members in 28 countries) and Friends of the Earth (FoE, with 0.5 million members in 46 countries).

The problem for all environmentalist groups is that they need to be constantly active, as threats to the environment are continuous. The groups therefore work on a variety of fronts at each level. They work with thinktanks such as the World Resources Institute (WRI) in Washington, DC, the Worldwatch Institute and the International Institute for Environment and Development (London). They also make use of networks such as the Earth Council (Maurice Strong and UNCED), the Centre for our Common Future, Geneva (Gro Harlem Brundtland and the former WCED) and Third World Network.

Environmental groups make up only some 10% of the 1000 NGOs with permanent UN affiliation. UNCED accredited 1400 itself and 1600 groups took part in Global Forum '92. NGOs had access to PrepCom meetings; they could attend (though not always simultaneously) all public sessions and they had the opportunity in the confined space of the convention centre to lobby national delegations. More controversially, 14 of the national delegations included some NGO representation, potentially an important precedent for other activities within the UN system. On the other hand, there was not so much evidence of NGO impact on existing specialized agencies, and particular disappointment at its effect (or lack of it) on the World Bank. By 1992 there was some concern at the emergence of 'fake NGOs' — bodies set up by governments to give them the sort of advice they wanted to hear.

International organizations want to deal directly with NGOs for five reasons. Publicity brings increased funding. The ideological preferences of such organizations are for consultation and deliberation. Friendly criticism and the ability to generate support both enhance programme effectiveness and sustainability. External pressure is more easily handled by dealing with organizations rather than with the public at large. Lastly, the creation of constituencies brings them support in the internal battle against other agencies and organizations (Uvin, 1996). In this way they can increase the legitimacy of the UN process. Conversely, of course, squaring NGOs beforehand also enables international organizations to 'capture' them and to defuse criticism of inadequate or even anti-environmental decisions. Nor can it be assumed that environmental NGOs are the only ones in the field. Other NGOs exist that are often better funded and press very different agendas: especially the energy lobby, whose activities were very much in evidence at Rio. Finger's criticisms cannot be ignored:

> Quite logically, the UNCED process sought to divide, co-opt, and weaken the green movement, a process for which the movement itself has some responsibility. On the one hand UNCED brought every possible NGO into the system of lobbying governments, while on the other it quietly promoted business to take over the solutions.
>
> (Finger, 1993, p. 36).

North–South cooperation has in the past been best where a clear common interest has existed. However, to get governments to take note of environmental issues requires publicity, and the need to target publicity on a mass audience can exacerbate strains between South and North. Campaigning works best where it sensationalizes issues. Even in literate AICs, the problem of 'scientific illiteracy' bemoaned by Dorothy Howell in 1992 remains, and without a dramatic message it is hard to gain public attention. Where Northern and Southern interests appear to clash, difficulties have rapidly multiplied: 'At UNCED, the gulf between the biggest environmental NGOs and many Southern NGOs in particular seemed almost as huge as between the respective governments on some (though not all) issues' (Grubb et al., 1993, p. 46).

In 1992, the EU proposed four principles that should govern decision-making on developmental projects of all kinds, and these apply equally in the South:

- The *precautionary principle* was that action to arrest the causes of environmental damage should not be delayed to wait for full scientific knowledge.
- The *principle of prior assessment* called, on the other hand, for full assessment of the risks before any activity likely significantly to damage the environment was allowed to proceed.

- The *principle that the polluter pays* focused on the need for individuals or companies polluting the environment to meet the public costs of cleaning it up.
- The *principle of non-discriminatory public participation* called for all people to be fully informed about potential interference with their environment and to have a right to have their views taken into account in the making of policy.

The World Commission, which first popularized the term 'sustainable development', had already taken the principle a stage further, and defined it in terms of two criteria: intra- and intergenerational equity. The next generation must be left a stock of quality assets, and not simply left to scrabble among the scraps.

Implementing these principles, however, requires strategic planning and action at international, national and local level in which the individual is actively involved. The problem is that although environmental problems are global, the mechanisms that exist to do something about them above all require planning and enforcement at national level. There is less difference in practice than in theory between those who believe in the priority of the state and those who support the notion of an international community (Keohane, 1986).

THE LIMITS OF GLOBALIZATION

Globalization can be defined as 'the process by which events, decisions, and activities in one part of the world can come to have significant consequences for individuals and communities in quite distant parts of the globe' (McGrew et al., 1992, p. 23). It is the key characteristic of the modern economic system and — some argue — of modernity itself, the goal towards which Southern states naturally aspire (Giddens, 1990). As a result, theorists of international relations have come increasingly to emphasize the systemic factors affecting state behaviour, rather than the individual decisions of states themselves.

The notion of systemic factors, however, implies the existence of an international system. By system we mean an enduring set of interactions between individuals, or, in this case, states. States are in themselves functional systems, and can be viewed either as such or as subsystems within the larger international context. Despite the formal absence of authority in the world system, states have in general behaved in an orderly way, which presupposes some notion of international order (Bull, 1977). The world is a very complex place, and with the speeding up of communications during the twentieth century we have all come to interact with one another, across national boundaries, to a much greater extent than was ever before possible. Hence, despite the formal absence of global political authority, states do act together cooperatively, with each other and with a variety of international organizations and NGOs, to make decisions that are for the most part effective. Indeed, in some areas, such as TV broadcasting or air traffic control, they have no practical alternative.

The impression of globalization has been reinforced by the strengthening of international economic organizations. These include the summit meetings of the Group of Seven, the IMF, the World Bank and, most recently, the WTO. With the end of the Cold War, G7 (and G8) summits have gained a new significance. The media correctly recognize that these are more important than sessions of the UN. Dr Henry Kissinger originally proposed a 'summit' to consider international economic issues in 1971. However, it was at the time of the first oil shock in 1974 that President Giscard of France, whose constitution gave him exceptional power among Western leaders, took up the idea as a way of concerting action among the leading capitalist economies, and it was at his invitation that the first meeting was held at Rambouillet in November 1975. 'The first essential feature of the summit was that it should be small, select and personal. It should be limited to countries which carried weight and influence. It should bring together those directly responsible for policy. They should be able to talk together frankly and without inhibitions' (Putnam and Bayne, 1987, p. 29). No Southern state qualifies for membership, but over time the agenda for the summit meetings has taken on board more and more global issues.

The summits of the G7 are just one sign of the increasing control by the First World of trade relationships between themselves and the South. The fact that it has recently been enlarged to include Russia says less about Russia's ambitions to be recognized as a First World country than about the fact that, as a result, it no longer poses a threat as an alternative centre of economic power. In the meantime, the conclusion of the Uruguay Round of the GATT has culminated in the formation of the WTO, which large companies are already using to eliminate competition to their interests.

At one time, the creation of regional economic groupings was seen in the South as the way both to encourage South–South trade and to defend themselves against the cartelization of the First World. However, the signing of the North American Free Trade Agreement (NAFTA), which further extends the already very large US internal market to Canada and Mexico, has already been followed by the signing of a free trade agreement with Chile and interest in joining the new cartel from the Caribbean states, which are, however, still being held at arm's length.

Today, with the Asian crisis following so soon on the 'tequila effect' on the rest of Latin America and the 1994–5 economic crisis in Mexico, Southern exports are stagnating. It is possible that 'generalized dumping' has been taking place (Bairoch, in Boyer and Drache, 1996, p. 178). It is difficult to tell, since the question of whether a price is fair is affected by differences between the strengths of currencies in selling and buying countries. Despite the strength of the yen, Japan continued to run a substantial positive trade balance with the rest of the world (Chorney, in Boyer and Drache, 1996). It is, however, not technology but politics that makes this possible, and this casts doubt on the universal assumption of rapidly increasing globalization. As Bienefeld presciently warned in 1996, 'The primary driving force behind the liberalization of the world's financial markets is political, not technological. The motive force has been the opportunity to amass untold fortunes through the creation of mountains of credit – and debt' (Bienefeld, in Boyer and Drache, 1996, pp. 420–1).

The one ray of light from the point of view of the South, in a world that seems increasingly dominated by the economic power of the North, is the past evidence of the North's consistent failure to understand that the existing situation can change. In 1945, it assumed the continued political dominance of then great powers, enshrined in permanent membership of the UN Security Council, and this no longer fits current realities. However, the structures remain unchanged. In theory the new institutions of economic power are different and should more easily adapt to the rise of new economic forces. It is true that the G7 has been reluctant to admit Russia, and there are very different views on China and the real strength of its economy: for rather different reasons both Bush and Clinton have tended to play it up. Forecasts of the continuing rise of Japan, on the other hand, now look overoptimistic. The crisis in South-East Asia in 1997, and the continuing problems presented for the Japanese economy by excessive lending to NICs such as South Korea, were not foreseen by authors such as Cole and Miles (1984), whose predictions about the world in 2010 already look wide of the mark.

WHO WILL PAY?

Other developed states, as well as Japan, continue to use funding as a lever of foreign policy. The 1980s was generally a period of movement towards a market ideology. The 1973–4 oil price rise had been successfully absorbed by the wealthier countries. But the 1978–9 rise produced recession in industrialized nations. This led in turn to a fall in commodity prices, increased protectionism, higher world interest rates, a scarcity of private loans and stagnating aid.

Developed states, too, are affected by the consequences of globalization: Five hundred corporations now control 70% of world trade, 80% of foreign investment and 30% of world GNP. The TNCs can afford to pay – but they can more easily use their wealth and influence to obtain greenfield sites, large subsidies, pleasant living conditions and bonuses for their directors. The responsibility for environmental protection remains at national level, but the problem is enforcing it. Poor countries cannot easily enforce the 'polluter pays' principle. They are fearful of trying to do so because the company concerned may simply shut down production and move elsewhere. Besides, since the 1980s, privatization has in many cases transferred responsibility for payment to customers, whose main interest is to keep the price they pay as low as possible. The main shareholders in these vast corporations, are themselves corporations and they show no real interest in enforcing higher standards of responsibility. It must therefore be the responsibility of the home states to control the actions of their TNCs abroad, and it is at least an encouraging sign that NGOs in the AICs have argued this and are prepared to use their influence on national governments, and, perhaps more importantly, on Northern consumers, to insist on certain minimum standards of behaviour.

Southern states suffered to a greater or lesser degree from the Debt Crisis of the 1980s. Many of them plead poverty and lack of resources as a reason for not taking any action on the environment, and environmentalists have not been slow to blame the IMF for forcing unreasonable terms on governments and so causing environmental damage (Adams, 1991). There are two special cases.

The governments of East and East Central Europe regard themselves generally as mature states capable of dealing with their own environmental problems. At Rio their leaders quite rightly sought to explain why their states were particularly important, on account of their peculiar terrain, unique mix of species or otherwise. The environmental arena was clearly one in which they could project their own independent national identity as players in the game of world politics. However, some of these states carry large debts and are finding it difficult to clean up the legacy of two generations of old-fashioned, unrestrained pillaging of economic resources. There are serious environmental problems in the former East Germany, Poland, the Czech Republic and Romania, and dealing with these is going to take a long time.

Some, mostly Middle Eastern, states, such as Saudi Arabia, Kuwait, the United Arab Emirates and Brunei, command a significant surplus of economic resources and are able to act as alternative sources of funding to Southern states. Otherwise, the main source of funds have been Western banks and international lending agencies, specifically the IMF and the World Bank.

When Southern countries turn to the IMF for help, the advice they receive in return is consistent, although their problems are not. The ideology of the IMF favours orthodox economic explanations of the need for stabilization. The Fund is an organization that is concerned primarily with the technicalities of maintaining the system. Consequently, if they want to borrow from the IMF, countries have to take measures that will as far as possible free up world trade, regardless of the consequences for themselves. They will be faced with the following specific requirements:

1. To reduce budget deficits through an immediate reduction in public expenditure.
2. To eliminate all forms of price and wage control, including the removal of subsidies on basic foodstuffs that often form an essential part of the bargain between Southern governments and key interests.
3. To control the money supply.
4. To devalue the currency in order to promote exports and reduce imports.
5. To remove tariffs and quotas that protect infant Southern industries.

These requirements are collectively referred to as Structural Adjustment Packages (SAPs). SAPs were (and are) based on the premise that an economy was fundamentally sound, and that what was needed was adjustment to a temporary balance-of-payments crisis. Placing emphasis on markets and exports at times of depressed commodity prices might seem suspect, however, and flooding the market will inevitably depress prices further. The result is bad for the environment. Southern states will give priority to what they can easily produce: primary products that sell at low prices on the world market. They have every incentive to use intensive agriculture to produce cash crops and to exhaust mineral resources as quickly as possible. SAPs advantage TNCs, as well as local elites who have business interests, by lowering wages and reducing the power of labour to demand a fairer return from employers.

Free market policies have not resolved the debt problem, although Southern governments have in the main successfully 'muddled through'. George (1992, p. 27) notes that despite paying out some US$1300 billion in debt repayments and interest in the period 1982–90, debtor countries were still 61% more indebted at the end of period than at the beginning. Worst of all, Africa South of the Sahara's debt had increased by 113%, although world interest rates had at last begun to fall back towards historically more typical levels. In the meantime trimming government spending has meant that in practice very low priority has been given to environmental issues, and the mechanisms for enforcement hardly exist.

The World Bank has since its foundation been seen as the main source of multilateral lending to countries for individual capital projects. It established the International Development Association in 1960 to give loans on easy terms to the poorest countries, and certainly since 1973 the World Bank has distinguished between relative and absolute forms of poverty at personal and national levels, has stressed investment in the poor and has funded projects concerned with small-scale production.

However, in the same year that the then President of the World Bank, Robert McNamara, said 'it is clear that too much confidence was placed on the belief that rapid economic growth would automatically result in the reduction of poverty' the 1977 World Bank Report on Africa said that aid should be given only where subsidies were abandoned, even if this meant food riots. And the World Bank has been repeatedly accused of supporting large-scale, technological 'solutions' (e.g. dams), to the detriment of the environment.

The evident failure of the IMF to deal with the problems of Southern debt has led the World Bank, too, to offer SAPs to specified countries. One of their attractions originally was that there was no cross-conditionality. For example, a World Bank loan to Argentina has been unsuccessfully opposed by the IMF. But during the 1980s an informal or tacit cross-conditionality became increasingly evident.

Meanwhile, the Bank had increasingly come under criticism from NGOs that regarded it as having acted consistently in sacrificing environmental goals in the search for economic growth.

World Bank support for colonisation projects such as the Polonoroeste project in Brazil and the transmigration project in Indonesia, which contributed to deforestation, came under fierce attack. A number of Bank-financed projects were detrimental to the environment. Bank finance for hydroelectric projects, for example, destroyed watersheds and flooded wildlife sanctuaries and portions of parks in Thailand, Malaysia, Brazil and Zaire. The Bank participated in cattle-ranching schemes in Latin America which were destructive of forested areas, and financed a cattle development in Botswana which contributed to desertification.

(Williams, 1994, p. 133).

In 1987, Barber Conable, in a speech to the World Resources Institute, acknowledged that the Bank had made serious environmental mistakes. He singled out the Brazilian Polonoroeste plan as what he termed 'a sobering example of an environmentally sound effort that went wrong' (Williams, 1994, p. 130). As a result, an Environment Department was created within the Bank to monitor proposals for their environmental effects and train officials in environmental awareness, and in 1990 a systematic screening of all new projects was introduced. Reed (1992, pp. 163–7) noted that some progress had already been made towards implementing four main recommendations stemming from this work:

1. 'Priority must be given to poverty-alleviating economic strategies with a particular focus on stabilizing and strengthening the agricultural sector' (p. 163).
2. SAPs must be based on principles 'safeguarding sustainable use of resources' (p. 164).
3. Lending should be directed towards rebuilding natural capital, restoring degraded natural resources and improving use of resources.
4. Expansion of trade should not be at the expense of the environment.

The World Bank administers the Global Environment Facility (GEF), a multibillion dollar fund to finance environmental projects in developing countries. It has four priority areas: global warming, loss of biological diversity, pollution of international waters and depletion of the ozone layer.

Since Rio, additional states have signed up to the Conventions on Biodiversity and Climate Change, but the way in which they are to be implemented remains complicated and the procedures are slow. In April 1992, an Eminent Persons Meeting in Tokyo decided that the Conventions should be funded by the GEF, and, following receipt of the Tokyo Declaration on Financing Global Environment and Development (United Nations, 1992), later in the same month 32 governments reached agreement on the restructuring of the GEF 'to ensure an equitable representation of the interests of developing countries while giving due weight to the funding efforts of donor countries'. Study of the projects listed in the first three tranches shows a wide distribution of funds, both geographically by region and country and across three of the four main areas to be covered: global warming, biodiversity and management of international waters. No projects were listed for the fourth area, ozone depletion. However, by June 1992 only a small selection, all but one from the first tranche, had actually been approved, examples being 'Emissions of global warming gases from rice soils', 'Protected areas and wildlife conservation in Vietnam' and 'Environmental management in the Danube river basin' (World Bank GEF, 1992 a, b; see also UNDP, 1994).

The main problems here are the extreme slowness of the bidding process and the need to distribute resources widely for political reasons. GEF projects were not intended as solutions to all ecological difficulties, but as pump-priming projects to encourage other contributions; meanwhile, national governments of the AICs have allowed their contribution in aid to drop thus making it even less likely that the demonstration effect will work. Projects to pay off debts of the very poorest nations have been bogged down in endless discussions. At the international level, therefore, there is a need to get general agreement to develop more effective, long-term mechanisms for multilateral funding.

STRATEGY OPTIONS

The reasons for concern about the future of the environment have only been strengthened by recent developments. Ozone depletion is now reaching high levels in the Antarctic, and there are clear signs of increasing depletion in the Arctic too. There is now much more evidence both of global warming and of the organized opposition in the USA and other AICs to establishing an effective limit on emissions that will come into effect soon enough to make any difference. The arguments of island and coastal states have simply been ignored.

The fires in Indonesia again covered South-East Asia with so-called 'haze' in 1998. The burning of Brazilian Amazonia has increased rather than decreased, and in Mexico, which has already lost 95% of its forest cover, a major forest fire was burning out of control on the border between Oaxaca and Chiapas. The collapse of the Newfoundland fisheries has been confirmed, and the vast majority of the world's fishing grounds have now reached critical levels.

Above all, it has been increasingly clear from the tone and content of international discussion that politicians still choose to ignore the need for sustainability, and that international organization is painfully slow and the strategy options open to national governments are very limited, even where the problem is actually understood. Where, as is generally the case, powerful lobbies oppose action, the issue of who will pay is dominant. To take the most obvious example, climate change is not a new issue. The long-term effects of the consumption of fossil fuels were predicted as long ago as 1896 (Arrhenius, 1896). The 'science' itself is not controversial; the problem is political. Aaron Wildavsky argues that the time has been taken up in debate between communitarians and advocates of the free market, but the wider issue is one of global governance (Rayner, 1995). Global action is only possible if the AICs, which control most of the votes in world political and economic institutions, are willing to allow it. Their track record since 1992 has not been impressive.

At Rio, it was the USA that constituted a major obstacle to progress towards sustainability. On two major issues, global climate change and biodiversity, the Bush administration was successful in ensuring that no effective action took place. But the USA was even at the time (1992) responsible for some 23% of world carbon dioxide emissions, and without some response from the USA, it was never likely that any effective action could take place. President Clinton won the 1992 election and initially made some positive environmental decisions, notably by signing the Biodiversity Convention. His Vice President, Al Gore, had a good record on environmental issues and is still regarded as the 'green' candidate for 2000. But Clinton has been ineffective and his record at both state and federal levels shows regular concessions to big business, most recently in leasing oil reserves in national park land and deferring dates for emissions reductions eventually agreed at Kyoto in December 1997.

Japan and the UK were among those AICs doing most to torpedo the Rio Conference. Both were and are very much seen as enthusiastic supporters of the USA. Japan under the Liberal Democrats was plagued by a series of short-lived ministries, which could not have taken action even if they had wanted to. Over the past four years, however, it has been sliding into recession, a decline accelerated by the Asian crisis of 1997. In 1992, the UK was out of step with the rest of Europe and only just emerging from recession. The Major government was able to make encouraging noises without the need to take action; Britain's carbon dioxide emissions were for several years below 1990 levels simply because of the poor state of the British economy. The new Prime Minister, Tony Blair, was generally seen as having made the most committed speech at UNGASS in New York in June 1997. He stressed then the redirection of UK aid to poverty and environmental problems, but despite the creation of an environmental superministry under the Deputy Prime Minister, John Prescott, there was no environmental legislation among the 26 Bills set out in the 1997 Queen's Speech.

WHAT IS THE PROBLEM?

Resource Crisis

It is true that there is no immediate resource crisis, in the sense that only a modest increase in the difficulty of recovering and hence in the price of any of the key raw materials is to be expected during the working lifetime of the politicians currently in office. However, although new sources may be found, mineral resources are finite. Energy constitutes the most critical problem. As noted in Chapter 2, petrochemicals are increasingly substituting for other minerals, especially metals. Oil and gas have limited availability already, so this means that their rate of depletion is likely to increase rather than decrease. Worse still, when they run out, there is no clear alternative.

Much confusion is regularly generated by the distinction between 'resources' and 'reserves'. *Reserves* are defined as known stocks of mineral ores and so on that can be exploited using current technology. On this basis, the reserves of minerals in 1989 included: iron ore 167 years (until 2156), aluminium 224 years (2213), copper 41 years (2030), silver, tin, zinc, mercury, lead 22 years (2011). Some substitution is possible (e.g. tin replaced by varnish on cans). Although new reserves have been found since 1950 and prices have been steadily falling, especially since the end of the Cold War, an increasing world population will deplete known reserves more quickly. The tin crisis in Bolivia is a sharp reminder of the knock-on effect of depletion on producer states in the South. Exhaustion of reserves could, as in this case, have serious social consequences in individual states even if it made no great difference to the world economy as a whole, since in the case of tin the collapse in the world price was a product both of increased supplies of cheap tin from Indonesia and Malaysia and declining demand in the AICs, as printed circuits replaced soldered joints, etc.

Mining interests defend current rates of depletion with the suggestion that a substitute will always be found. This would be a more convincing argument if there were not weighty evidence against.

First, the direction of causation is not fixed: there is convincing historical evidence that availability is itself one of the triggers to technological innovation. Second, by definition a substitute will always be more expensive, at least to begin with. Third, there may be technical problems that cannot be overcome. For example, a number of synthetics, such as fibreglass, can be used as a substitute for steel in the manufacture of cars, but fibreglass resins draw on petrochemicals and in addition are highly polluting to the environment. In any case, the advanced technology required to substitute for many of the common metals requires a degree of technological sophistication unlikely at present to be available in a Southern state. Hence, Southern governments insist on including provisions on the transfer of technology in any contract with a TNC.

There are two main inventions which both act to make the resource crisis worse and discourage any effective action to forestall it: cars and television. If the South had as many cars per person as the North does now there would be eight times as many in the world. Possible changes which would make the car less damaging are conservation of energy, better public transport, alternative energy sources and recycling. So far, only the first of these is being pursued by manufacturers with any enthusiasm, since economy has proved to be a selling point. However, US pressure to produce a non-polluting vehicle has yet to bear fruit, and even an electric car requires the consumption of a great deal of energy to recharge it. Recycling occurs to some extent but so far has had a rather limited effect.

Television's contribution to the resource crisis is an indirect one. It is a major agent of consumerism, encouraging all of us to aspire to Hollywood standards, which even most Americans cannot hope to attain. Governments are reluctant to try to check it. They fear the accusation of censorship, and those governments that do not care about such a charge are usually concerned only with matters of sexual morality and see nothing wrong with consumption of the right sort of products. Where such governments resist the impact of global television images their success is likely to be short-lived. Iranian satellite dishes may be hidden below the parapets of houses, but they exist and are now generally ignored by the authorities. CNN and MTV carry images and advertisements worldwide, as the number of satellite dishes in Brazilian *favelas* attest.

Cultural aspects of globalization may be unstoppable, but there needs above all to be action at national level by state governments to economize in the use of their 'own' resources, and to plan wisely to 'harvest' slowly and reinvest proceeds in alternative forms of development. A good example is the prudent use of Venezuelan oil between 1960 and 1985.

Land Crisis

High land yield involves high energy input, and is unsustainable. In any case, countries such as India and Bangladesh are already farming 97% of the suitable land. However, living standards have to be raised for 1.1 billion of the world's population.

Degradation and soil erosion can be controlled, although it is much harder to reduce them to a level that ensures the possibility of sustainability. Wind and water are the main agents, and in the absence of positive measures will erode fragile lands. Human factors are crucial. The planting of windbreaks, hedges and forests, the use of contour farming and livestock destocking are technical means to sustainability, and with proper husbandry it has already been demonstrated that land degradation can be halted and even reversed.

On the other hand, social structures can actively accelerate land degradation. In Zimbabwe, the most vulnerable land, significantly, has not been eroded. Eighty-three per cent of erosion has occurred on the former communal lands as a result of population growth, while the large estates seem relatively unaffected. In Fiji, on the other hand, the fact that Indo-Fijians cannot own land (which has to be rented from Fijians) gives them no incentive to conserve the land they farm (Clarke and Morrison, 1987).

Policy decisions are important too. Deforestation has been accelerated by subsidies paid to farmers in Indonesia. Ending subsidies would begin the move back to sustainability. However, in Amazonia there is no evidence yet that land cleared in the 1970s will regenerate from heathland of its own accord. Crop rotation, intercropping with legumes, selective planting of trees and mulching with leaf litter all help, but they need strong encouragement and some investment by government. Again, action is needed at national level. Action to end harmful subsidies would again be a first step, but it would then be necessary to educate farmers to implement sustainable farming, with appropriate financial support to ease the transition.

Deforestation involves the additional problem of predatory logging companies. In 1989, less than 0.1% of tropical logging was being done on a sustainable basis. However, since most of the companies concerned are foreign-owned, it should be easier rather than harder for action to be taken on a national level to restrain their depredations, were it not for the corrupt deals that are done between the companies and powerful local interests. But there are some signs of the necessary action. 'Glimpses of a "new forestry" have appeared in recent years, an approach that views forests as complex ecosystems providing a multitude of benefits, rather than simply as suppliers of lumber and fibre. Logging is carried on in a manner that minimises destruction and enables the forests' rapid recovery' (Brown *et al.*, 1992, p. 76). Individual countries could develop a policy on forests with both clear inducements and effective enforcement. There is also a need to minimize waste, such as the burning of pallets, and to step up tree planting in the AICs, both to save tropical timber and to demonstrate that the North is making its own efforts to counter global warming.

Pollution Crisis

The crisis of pollution and degradation has been slower to develop. It has for much of the recent period had only localized effects in the South, and the need to cope with it is not seen as a high priority by Southern ruling elites. It has usually not been much of a priority for Northern governments either.

As a transboundary problem, the pollution of the atmosphere needs international action. The Montreal Convention ending the production and use of CFCs was a relative success for international action because the scientific argument was unusually clear (so much so that it was accepted by Mrs Thatcher) and there was relatively little at stake – the manufacture of CFCs was not a major activity for any chemical industry and represents only a minute fraction of the output of a modern industrial economy. There was and is no such unanimity on carbon dioxide emissions, and in the Arab States of the Middle East a very powerful lobby exists in favour of burning fossil fuels. Generally speaking, moreover, developing countries do not see pollution, they see growth.

The 'polluter pays' principle was laid down in Brundtland but honoured only in principle at Rio. In the developed countries between 1973 and 1990 there was a slow movement to deal with those aspects which were seen as most immediate, whether or not they were the most important. Thus, in the UK the use of lead-free petrol was given fiscal encouragement, and a landfill tax was introduced, but the UK met targets on gaseous emissions only fortuitously. National self-interest was otherwise invoked to prevent international action. A real move away from the petroleum-rich economy of the AICs involves the active cooperation of the 'Seven Sisters' (see Chapter 2) and real international cooperation on production targets. However, OPEC has consistently proved unable to enforce production quotas even when it commanded the bulk of the oil market and the producer states of the South saw themselves as having a common interest in doing so. Consumers (including the US government) invoked economy measures for only the briefest of periods before taking concerted action to break down producer unity. And they were aided in this by the evident lack of concern that the oil states showed towards the plight of the rest of the South, hit as they were by both rising oil prices and a decline in demand for their products from the North.

The major generalized source of atmospheric pollution is motor transport. Rigorous controls on emissions from internal combustion and diesel engines have been enacted by some Northern states, but are rarely enforced effectively; in the NICs the cruder alternative has been tried of limiting urban car use by the selective use of number plates. A big problem for Southern states will be the high capital cost of new vehicles capable of meeting more stringent emission limits. Fitting scrubbers to thermal power stations would be the most effective measure that could realistically be adopted. The combined effect of both would have an immediate impact on the problem of photochemical smog, and a longer-term effect in countering the acidification of the atmosphere and wet deposition in rivers and lakes. However, in the case of acidification, because of the boundary problem international action will also be essential and, given the weakness of Southern regional organizations, it would be expecting too much to imagine that any decisions made could be effectively implemented.

In the South, international action has had little effect: the law of the sea needs enforceable provisions and the North has been able to ignore the wishes of the South and to continue in effect to enclose the commons. On liquid pollution, the EU has taken a lead in enforcing standards against the wishes of its national governments. In the South, the recycling of solid wastes is a major industry, and this is one area in which the South has proved more efficient than the North. The use of expensive excessive packaging is a recent phenomenon there, anyway. However, control of dumping is often very lax. Effective provision for such recycling has to be local, laws against dumping must be enacted at national level and both will have to confront the interests of both Northern governments and, more importantly, Northern TNCs.

WHAT CAN BE DONE AT THE NATIONAL LEVEL?

At the national level ecotaxes could in principle be imposed or raised.

Environmentalists argue that if the polluter were really to pay the full social and ecological costs, these would of course be passed on to the consumer, but this would at least make more environmentally damaging products more expensive than less damaging ones. This in turn should stimulate production of less damaging products and stimulate employment, since human labour is likely again to become more economic in comparison with mechanized production. The problem is competition from cheap goods produced where regulation is weakest.

Worse still, the creation of the WTO is already having precisely the wrong effect through its ideological orientation towards 'free' trade. Its decision that EU policies to maintain smaller-scale, more environmentally friendly banana production in the Caribbean islands are contrary to its rules, and must be ended, followed a complaint from the giant US grower Chiquita, which already has 70% of the EU market (The Guardian, 25 September 1997).

Management of waste has already had to be tackled at national/local level. Recycling brings enormous gains and limits the need for further mining (Brown et al., 1992, pp. 65–6). Recycling can be made economically more attractive than disposal. But it is a fiddly process, there have to be special facilities to make it work and the design of products must take the need to recycle into account if the full benefits of recycling are to be realized (e.g. a requirement for car manufacturers to make cars more recyclable). And with globalized production, international efforts are required to secure this.

Landfill cannot be allowed indefinitely. Already in the USA leakage is leading to escalating insurance claims, and the generation of methane has many objections. Taxing landfill is a beginning; however, incineration enhances global warming too. Again, 'solving' a local/national problem may exacerbate a global one. Pollution of the ocean bed is a growing problem off industrialized coasts, and so is the inability or unwillingness to regulate overfishing.

All these examples make clear the need for government action to regulate industrial production. The only acceptable solution in the long term must be closed-cycle processing. In other words, all industrial processes must recycle the materials they use (though there will be, of course, some cost, possibly a substantial one, in terms of energy input). Northern companies and international organizations must take the lead in establishing higher environmental standards worldwide. If they do not, the free flow of goods from areas without strict regulation will undercut producers who abide by national environmental standards, and drive them out of business.

THE PROBLEM OF POVERTY

Governments have the prime role to play in tackling the inequality that encourages the rich to be wasteful and the poor to overexploit their environment. Large-scale technology must continue to be used for the foreseeable future: for example, modern sanitation is essential if cities are not to become unsafe. Governments must have a strategy for health, education and development, even if they find it hard to implement.

Ill-health is expensive and wasteful. Unsafe water and malnutrition are major causes of continuing ill-health and early death. The first target has been improving water supply and sanitation, and in the case of the former there are some real success stories. However, with expanding populations and urbanization an even greater effort has to be made over the next 20 years. In many countries in Latin America, Asia and North Africa land reform is vital if people are to be fed properly and work is to be made available. Neoliberal ideology runs counter to land reform, and this will be difficult to achieve, despite the fact that successful land reform and the equalization of wealth is now generally acknowledged to have been a key factor in the spectacular growth of both Japan and South Korea since 1945. Profiting from their mistakes, other Southern states can take advantage of greater knowledge to avoid some of the environmental mistakes that were made at the same time.

Education of women is recognized to be a main factor in halting population growth. Women are also of key importance in strategic decisions about how households consume. Children have to be educated to see how the world can be conserved, beginning with informal socialization in the family and the primary curriculum. Education teaching youngsters about the threat inherent in consumerism will be vital to developing their ability to resist the demands of fashion.

No world can possibly be sustainable when the majority are denied the opportunity to enjoy the benefits of development already commonplace to the well-to-do. There is an overwhelming need to prioritize the provision of what the Prince of Wales in 1992 termed 'the five great liberators of development' (Prins, 1994, p. 11):

- cookers (which conserve fuelwood resources);
- refrigerators (which keep food from spoiling);
- water pumps (which avoid diseases associated with scarce surface water supplies and save so much female time);

- radios (which provide cheap links to the outside world);
- lights (which extend the working day and improve recreational possibilities).

Over much of the earth's surface it is now possible to provide solar power locally where it is needed, and so to end dependence on fossil fuel consumption. The problem is that to be effective the manufacture and distribution of non-polluting products will have to be subsidized, at least in the first instance. Governments in the South can encourage these, but China's example reminds us that development still takes priority over the environment: the manufacture of refrigerators in China, despite its obvious benefits from a health point of view, causes particular anxiety about the continuing use and handling of CFCs. At international level, therefore, agreement is needed to prevent futile competition and hence waste; the exact opposite of the principles followed by the WTO and the neoliberal world of the 1990s.

CONCLUSION

Throughout we have repeatedly observed that certain themes recur again and again in the environmental politics of Southern states. The underlying message, however, is that we live in one world and that the problems of the South, which are only likely to get worse, require action at both national (First World and Third World) and international level. Unfortunately, in the current international political climate, improvements in international cooperation do not seem very likely. Time and again we come back to the fact that the environment is global but the world political structure requires that responses to it are still enforced nationally.

The optimistic scenario is that the risks of global climate change, growing water shortage and rising levels of pollution have been exaggerated, and that as a result the interminably slow processes of international agreement will work at least fast enough to avert a further deterioration in the environment from 2040 onwards. The pessimistic scenario is that, by that stage, not only will such serious losses have occurred among irreplaceable plant and animal species that the world of the future will be much less attractive to live in, but the growing instability of the biosphere may result in some way in a catastrophic collapse of the human population, possibly as early as the time of our children's children. The limits of adaptation to climate change, rise in sea level, etc, have already been reached.

Combatting these problems is going to take a great deal of money. The value to the AICs of strong Southern economies seems to be widely accepted. The political response in terms of rising taxes and heavy spending on long-term measures, however, will only come if Northern politicians stop taking the easy options. In the end, both North and South have a common interest in arresting environmental degradation. At Rio in 1992, the Minister of Finance of the Republic of Nauru described how mining for phosphate had left one-third of his tiny country a waste of dead coral, leaving its 5000 population crowded into the part that remained. Although in time the damage to his island could be restored, that to the planet could not. It was, he said, an 'omen to all the people of Earth'.

BIBLIOGRAPHY

Abu-Lughod, J. and Jay, R. Jr (eds) (1977) *Third World Urbanization*. London: Methuen.

Adams, P. (1991) *Odious Debts: Loose Lending, Corruption and the Third World's Environmental Legacy*. London: Earthscan.

Agarwal, B. (1986) *Cold Hearths and Barren Slopes: the Woodfuel Crisis in the Third World*. London: Zed Books.

Albert (1994) In B. R. Johnston (ed.), *Who Pays the Price?* Island Press.

Alexis, L. (1983) The damnation of paradise: Sri Lanka's Mahaweli scheme. Unpublished paper.

Allison, G. T. (1971) *Essence of Decision: Explaining the Cuban Missile Crisis*. Boston, MA: Little Brown.

Andrews, J. E., Brimblecombe, P., Jickells, T. D. and Liss, P. S. (1996) *An Introduction to Environmental Chemistry*. Oxford: Blackwell Science.

Arecchi, A. (1985) Dakar, *Cities*, 2(3), 198–211.

Armstrong, W. and McGee, T. G. (1985) *Theatres of Accumulation: Studies in Asian and Latin American Urbanization*. London: Methuen.

Arnon, I. (1981) *Modernizing Agriculture in Developing Countries: Resources, Potential and Problems*. Chichester: Wiley.

Arrhenius, S. (1896) On the influence of carbonic acid in the air upon the temperature of the ground. *Philosophical Magazine*, S5, 41(251).

Barbier, E. B. and Markandya, A. (1989) *The Conditions for Achieving Environmentally Sustainable Development*. London: International Institute for Environment and Development.

Barkham, J. (1995) Ecosystem Management and environmental ethics. In T. O'Riordan (ed.), *Environmental Science for Environmental Management*. Harlow: Longman.

Barrow, C. J. (1995) *Developing the Environment: Problems and Management*. Harlow: Longman.

Bell, D. (1987) The world and the United States in 2013. *Daedalus*, 116, 3.

Blaikie, P. (1985) *The Political Economy of Soil Erosion in Developing Countries*. London: Longman.

Blaikie, P. and Brookfield, H. (1987) *Land Degradation and Society*. London: Methuen.

Bogard, W. (1989) *The Bhopal Tragedy: Language, Logic and Politics in the Production of a Hazard*. Boulder, CO: Westview Press.

Boserup, E. (1965) *The Conditions of Agricultural Growth*. London: Allen & Unwin.

Boserup, E. (1983) The impact of scarcity and plenty on development. In R.

Rotberg and T. Rabb (eds), *Hunger and History*. Cambridge: Cambridge University Press.

Bowonder, B., Kasperson, J. X. and Kasperson, R. E. (1985) Avoiding future Bhopals. *Environment*, September, 10.

Boyer, R. and Drache, D. (1996) *States against Markets: the Limits of Globalization*. London: Routledge.

Bradbury, I. (1991) *The Biosphere*. London: Belhaven Press.

Brammer, H. (1990) Floods in Bangladesh. II: Flood mitigation and environmental aspects. *Geographical Journal*, 156(2), 158–65.

Brandt, W. (1980) *North–South: a Programme for Survival. Report of the Independent Commission on International Development Issues (The Brandt Commission)*. London: Pan Books.

Brasil, Governo do (1992) *The challenge of sustainable development: the Brazilian Report for the United Nations Conference on Environment and Development*. Brasilia, Press Secretariat of the Presidency of the Republic.

Brasil, Governo do (1992) *Ethanol: Brazil's cleaner fuel*. Brasilia, n.p.

Brazil AIAA/SOPRAL (1992) *Ethanol: Energy Source for a Sustainable Society*. Rio de Janeiro: AIAA/SOPRAL.

Browder, J. O. (ed.) (1989) *Fragile Lands of Latin America: Strategies for Sustainable Development*. Boulder, CO: Westview with Roger Thayer Stone Center for Latin American Studies of Tulane University.

Brown, L. R. (1981) *Building a Sustainable Society*. London: Norton.

Brown, L. R., Flavin, C. and Postel, S. (1993) *Saving the Planet: How to Shape an Environmentally Sustainable Global Economy*. London: Earthscan.

Brown, L. R. and Kane, H. (1995) *Full House: Reassessing the Earth's Population Carrying Capacity*. London: Earthscan.

Brown, P. (1997) Man's greed fuels global bonfire. *The Guardian*, 17 December.

Bull, H. (1977) *The Anarchical Society*. London: Macmillan.

Bunker, S. (1985) *Underdeveloping the Amazon: extraction, unequal exchange and the failure of the modern state*. Urbana, IL, University of Illinois Press.

Calvert, P. and Reader, M. (1998) Water resource management in Brazil. In P. O. Agbese and D. Vajpeyi (eds), *Water Resource Management in a Comparative Perspective*. Westport, CT: Greenwood Press.

Carson, R. (1962) *Silent Spring*. New York: Fawcett Crest.

Carter, A. P. (ed.) (1976) *Energy and the Environment: a Structural Analysis*. Hanover, NH: University Press of New England for Brandeis University Press.

Castells, M. (1977) *The Urban Question*. London: Edward Arnold.

Cathcart, R. B. (1983) Mediterranean Basin–Sahara reclamation. *Speculations in Science and Technology*, 6, 150–2.

Cherfas, J. (1988) *The Hunting of the Whale*. London: Bodley Head.

Chilcote, R. H. (1990) *Power and the Ruling Classes in Northeast Brazil: Juazeiro and Petrolina in Transition*. Cambridge: Cambridge University Press.

Choucri, N. with Ferraro, V. (1976) *International Politics of Energy*

Interdependence: the Case of Petroleum. Lexington, MA: D. C. Heath (Lexington Books).

Clarke, W. and Morrison, J. (1987) Land mismanagement and the development imperative in Fiji. In P. Blaikie and H. Brookfield (eds), *Land Degradation and Society*. London: Methuen.

Cleary, D. (1991) The greening of the Amazon. In D. Goodman and M. Redclift (eds), *Environment and Development in Latin America: the Problem of Sustainability*. Manchester: Manchester University Press.

Cole, J. (1987) *Development and Underdevelopment: a Profile of the Third World*. London: Methuen.

Cole, S. and Miles, I. (1984) *Worlds Apart: Technology and North–South Relations in the Global Economy*. Brighton: Wheatsheaf for UNITAR.

Collor de Mello, F. (1992) *Agenda for Consensus: a Social-liberal Proposal*. Brasilia: Governo do Brasil.

Conca, K. (1996) Greening the UN: environmental organizations and the UN system. In T. G. Weiss and L. Gordenker (eds), *NGOs, the UN, and Global Governance*. Boulder, CO: Lynne Reinner.

Cornelius, W. A. Jr (1971) The political sociology of cityward migration in Latin America: toward empirical theory. Reprinted in J. Abu-Lughod and R. Jay Jr (eds, 1977), *Third World Urbanization*. London: Methuen.

Crabtree, J., Duffy, G. and Pearce, J. (1987) *The Great Tin Crash: Bolivia and the World Tin Market*. London: Latin American Bureau.

Crenson, M. (1971) *The Un-politics of Air Pollution: a Study of Non-decision Making in the Cities*. Baltimore, MD: Johns Hopkins University Press.

Cummings, B. J. (1990) *Dam the Rivers: Damn the People. Development and Resistance in Amazonian Brazil*. London: Earthscan.

Cummings, R. G. (1974) *Interbasin Water Transfers: a Case Study in Mexico*. Baltimore, MD: Johns Hopkins University Press.

Dahl, R. (1961) *Who Governs? Democracy and Power in an American City*. New Haven, CT: Yale University Press.

Dando, W. A. (1980) *The Geography of Famine*. London: V. H. Winston & Sons.

Decker, R. W. and Barbara, B. (1991) *Mountains of Fire: the Nature of Volcanoes*. Cambridge: Cambridge University Press.

De los Reyes, P. J. (1992) Volunteer observers program: a tool for monitoring volcanic and seismic events in the Philippines. In G. J. H. McCall, D. J. C. Laming and S. C. Scott (eds), *Geohazards: Natural and Man-made*. London: Chapman & Hall.

de Oliveira, O. and Roberts, B. (1996) Urban development and social inequality in Latin America. In J. Gugler (ed.), *The Urban Transformation of the Developing World*. Oxford: Oxford University Press.

Desai, M. and Redfern, P. (eds) (1995) *Global Governance: Ethics and Economics of the World Order*. London: Pinter.

Devereux, S. (1993) *Theories of Famine*. Hemel Hempstead: Harvester Wheatsheaf.

Dickenson, J. P. et al., (1983) *A Geography of the Third World.* London: Methuen.

Drew, E. (1981) *Portrait of an Election: the 1980 Presidential Campaign.* London: Routledge & Kegan Paul.

Dryzek, J. (1997) *The Politics of the Earth: Environmental Discourses.* Oxford: Oxford University Press.

Earl, D. E. (1975) *Forest Energy and Economic Development.* Oxford: Clarendon Press.

Ecologist (1993) *Whose Common Future? Reclaiming the Commons.* London: Earthscan.

Ehrlich, P. (1968) *The Population Bomb.* New York: Ballantine Books.

Eisenstadt, S. (1963) *The Political Systems of Empires.* New York: The Free Press.

El Azhary, M. S. (ed.) (1984) *The Impact of Oil Revenues on Arab Gulf Development.* London: Croom Helm with Centre for Arab Gulf Studies, University of Exeter, and the Petroleum Information Committee for the Arab Gulf states.

El-Gawhary, K. (1995) Delta blues: the Nile. *New Internationalist,* November.

Elian, G. (1979) *The Principle of Sovereignty over Natural Resources.* Alphen aan den Rijn, Netherlands: Sijthoff and Noordhoff.

Elsom, D. M. (1992) *Atmospheric Pollution: a Global Problem.* Oxford: Blackwell.

Epstein, P. et al., (1996) *Current Effects of Climate Change.* Washington, DC: Ozone Action Roundtable (curreff2.html).

Esteva, G. (1992) Development. In W. Sachs (ed.), *The Development Dictionary: a Guide to Knowledge as Power.* London: Zed Books.

Finger, M. (1993) Politics of the UNCED Process. In W. Sachs (ed.), *Global Ecology: a New Arena of Political Conflict.* London: Zed Books.

Forbes, D. K. (1984) *The Geography of Underdevelopment: a Critical Survey.* London: Croom Helm.

Freedman, L. and Karsh, E. (1994) *The Gulf Conflict.* London: Faber & Faber.

Frjeka, T. (1994) Long-range global population projections: lessons learned. In W. Lutz (ed.), *The Future Population of the World: What Can We Assume Today?* London: Earthscan with International Institute for Applied Systems Analysis.

Garner, R. (1996) *Environmental Politics.* London: Prentice Hall/Harvester Wheatsheaf.

George, S. (1992) *The Debt Boomerang: How Third World Debt Harms Us All.* Boulder, CO: Westview Press.

Giddens, A. (1990) *The Consequences of Modernity.* Cambridge: Polity Press.

Gilbert, A. G. (1976) The argument for very large cities reconsidered. *Urban Studies,* 13, 27–34.

Gilbert, A. (1994) *The Latin American City.* London: Latin American Bureau.

Glasbergen, P. and Blowers, A. (1995) *Environmental Policy in an International Context: Perspectives.* Milton Keynes: Open University.

Goodland, R., Daly, H., El Serafy, S. and von Droste, B. (eds) (1991)

Environmentally Sustainable Economic Development: Building on Brundtland. Paris: UNESCO.

Goodman, D. and Hall, A. (1990) *The future of Amazonia.* New York, St Martin's Press.

Goudie, A. (1986) *The Human Impact on the Natural Environment,* 2nd edn. Oxford: Blackwell.

Grubb, M., Koch, M., Munson, A., Sullivan, F. and Thomson, K. (1993) *The Earth Summit Agreements: a Guide and Assessment.* London: Earthscan.

Gugler, J. (ed.) (1996) *The Urban Transformation of the Developing World.* Oxford: Oxford University Press.

Guimarães, R. P. (1991) *The Ecopolitics of Development in the Third World: Politics and Environment in Brazil.* Boulder, CO, and London: Lynne Rienner.

Haas, P. M., Keohane, R. O. and Levy, M. A. (eds) (1993) *Institutions for the Earth: Sources of Effective International Environmental Protection.* Cambridge, MA: MIT Press.

Hall, A. L. (1989) *Developing Amazonia: Deforestation and Social Conflict in Brazil's Caralds Programme.* Manchester: Manchester University Press.

Hall, M. L. (1992) The 1985 Nevado del Ruiz eruption: scientific, social and governmental response and interaction before the event. In G. J. H. McCall, D. J. C. Laming and S. C. Scott (eds), *Geohazards: Natural and Manmade.* London: Chapman & Hall.

Hallwood, P. and Sinclair, S. (1981) *Oil, Debt and Development: OPEC in the Third World.* London: Allen & Unwin.

Hanlon, J. (1986) *Beggar Your Neighbours: Apartheid Power in South Africa.* London: Catholic Institute for International Relations and James Carey Ltd.

Hansson, L. (ed.) (1992) *Ecological Principles of Nature Conservation: Applications in Temperate and Boreal Environments.* London: Elsevier Applied Science.

Hardoy, J. E. and Satterthwaite, D. (eds) (1986) *Small and Intermediate Urban Centres: Their Role in National and Regional Development in the Third World.* London: Hodder & Stoughton.

Harris, N. (1986) *The End of the Third World? Newly Industrialising Countries and the Decline of an Ideology.* London: I. B. Tauris.

Harrison, P. (1992) *The Third Revolution: Population, Environment and a Sustainable World.* Harmondsworth: Penguin.

Hepple, L. W. (1986) Geopolitics, generals and the state in Brazil. *Political Geography Quarterly,* 5, Supplement, 579–90.

Hertsgaard, M. (1989) *On Bended Knee: the Press and the Reagan Presidency.* New York: Schocken Books.

Hewitt, C. J. (1974) Elite and the distribution of power in British society. In P. Stanworth and A. Giddens (eds), *Elites and Power in British Society.* Cambridge: Cambridge University Press.

Hill, P. and Vielvoye, R. (1974) *Energy in Crisis: a Guide to World Oil Supply and Demand and Alternative Resources.* London: Robert Yeatman.

Hirsch, F. (1977) *The Social Limits to Growth.* London: Routledge & Kegan Paul.

Holdgate, M. W., Kassas, M. and White, G. F. (eds) (1982) *The World Environment 1972–1982: a Report by the United Nations Environment Programme*. Dublin: Tycooly International.

Hollis, G. E. (1978) The falling levels of the Aral and Caspian Seas. *Geographical Journal*, 144, 62–80.

Howell, D. J. (1992) *Scientific Literacy and Environmental Policy: the Missing Prerequisite for Sound Decision-making*. New York: Quorum Books.

Hugo, G. (1996) Urbanization in Indonesia. In J. Gugler (ed.), *The Urban Transformation of the Developing World*. Oxford: Oxford University Press.

Hurst, P. (1990) *Rainforest Politics: Ecological Destruction in South-East Asia*. London: Zed Books.

Ince, M. (1990) *The Rising Seas*. London: Earthscan with Commonwealth Secretariat.

IUCN (1996) *Red List of Threatened Animals*. Gland, Switzerland: IUCN.

IUCN/UNDP/WWF (1980) *World Conservation Strategy: Living Resource Conservation for Sustainable Development*. Gland, Switzerland: IUCN/UNDP/WWF.

Jenkins, R. (1987) *Transnational Corporations and Uneven Development: the Internationalization of Capital and the Third World*. London: Routledge.

Jones, G. E. (1987) *The Conservation of Ecosystems and Species*. London: Croom Helm.

Keohane, R. O. (ed.) (1986) *Neorealism and Its Critics*. New York: Columbia University Press.

Keohane, R. and Nye, J. (eds) (1977) *Transnationalism and World Politics*, 2nd edn. Boston: Little Brown.

Keesing's Record of World Events, various years.

Kliot, N. (1994) *Water Resources and Conflict in the Middle East*. London: Routledge.

Krasner, S. D. (ed.) (1983) *International Regimes*. Ithaca, NY: Cornell University Press.

Kubursi, A. A. (1985) Industrialisation: a Ruhr without water. In T. Niblock and R. Lawless (eds), *Prospects for the World Oil Industry*. London: Croom Helm.

Lappé, F. M. and Schurman, R. (1989) *Taking Population Seriously*. London: Earthscan.

Leatherman, S. P. (1995) *1995 Intergovernmental Panel on Climate Change Second Assessment Report*. Department of Natural Resources Protection, Broward County, quoted by Ozone Action Fact Sheet (p20.html).

Leroy, M. (1978) *Population and World Politics*. Lieden: Martinus Nijhoff.

Lukes, S. (1986) *Power*. Oxford: Blackwell.

Lutz, W. (ed.) (1994) *The Future Population of the World: What Can We Assume Today?* London: Earthscan with International Institute for Applied Systems Analysis.

Lutz, W., Prinz, C. and Langgassner, J. (1994) The IIASA world population scenarios to 2030. In W. Lutz (ed.), *The Future Population of the World: What*

Can We Assume Today? London: Earthscan with International Institute for Applied Systems Analysis.

Lyster, S. (1985) *International Wildlife Law*. Cambridge: Grotius.

McAuslan, P. (1985) *Urban Land and Shelter for the Poor*. London: Earthscan.

McCall, G. J. H., Laming, D. J. C. and Scott, S. C. (eds) (1992) *Geohazards: Natural and Manmade*. London: Chapman & Hall.

McCormick, J. (1989) *Reclaiming paradise: the global environmental movement*. Bloomington, IN, Indiana University Press.

McCully, P. (1996) *Silenced Rivers: the Ecology and Politics of Large Dams*. London: Zed Books in association with The Ecologist and International Rivers Network.

McCully, P. (1997) Earthquake hits Narmada Valley. *Third World Resurgence*, 84, 4.

McGrew, A. G. Lewis, P. G. *et al.*, (1992) *Global Politics: Globalization and the Nation-state*. Cambridge: Polity Press.

Main, H. A. C. (1990) Housing problems and squatting solutions in metropolitan Kano. In R. B. Potter and A. T. Salau (eds), *Cities and Development in the Third World*. London: Mansell.

Mannion, A. M. (1991) *Global Environmental Change*. Harlow: Longman.

Maya Institute (1978) *The Mahaweli Project*. Colombo: Maya Institute.

Meek, J. (1997) China joins scramble for black gold. *The Guardian*, 29 September.

Mehta, G. (1993) *A River Sutra*. London: Heinemann.

Middleton, N., O'Keefe, P. and Moyo, S. (1993) *Tears of the Crocodile: from Rio to Reality in the Developing World*. London: Pluto Press.

Moffett, G. D. (1994) *Critical Masses: the Global Population Challenge*. Harmondsworth: Penguin.

Morales, S. C. (1978) *The Off-shore Petroleum Resources of South-East Asia: Potential Conflict Situations and Related Economic Considerations*. Kuala Lumpur: Oxford University Press.

Morgan, R. P. C. (1986) *Soil Erosion and Conservation*. Harlow: Longman.

Mumford, J. L. *et al.*, (1987) Lung cancer and indoor pollution in Xuan Wei, China. *Science*, 235, 217–35.

Murphey, R. (1996) History of the city in monsoon Asia, in J. Gugler (ed.), *The Urban Transformation of the Developing World*. Oxford: Oxford University Press.

Myers, N. (ed.) (1987) *The Gaia Atlas of Planet Management for Today's Caretakers of Tomorrow's World*. London: Pan Books

Myers, N. (1991) *Population, Resources and the Environment: the critical challenges*. New York: UNFPA.

Neale, G. (1996) Running on empty. *The Sunday Telegraph*, 25 February.

Newson, M. (1992) *Land, Water and Development: River Basin Systems and Their Sustainable Management*. London: Routledge.

Niblock, T. and Lawless, R. (eds) (1985) *Prospects for the World Oil Industry*. London: Croom Helm.

Nordlinger, E. A. (1977) *Soldiers in Politics: Military Coups and Governments*. Englewood Cliffs, NJ: Prentice Hall.

Noreng, O. (1978) *Oil Politics in the 1980s: Patterns of International Cooperation*. New York: McGraw-Hill for the Council on Foreign Relations.

Nunn, P. (1997) Global warming. *New Internationalist*, June.

O'Donnell, G. (1988) *Bureaucratic-authoritarianism: Argentina, 1966–1973, in Comparative Perspective*. Berkeley: University of California Press.

Onoh, J. K. (1983) *The Nigerian Oil Economy: from Prosperity to Glut*. London: Croom Helm.

O'Riordan, T. (ed.) (1995) *Environmental Science for Environmental Management*. Harlow: Longman.

Panorama (1993) *Arming for Islam*. London: BBC Television.

Patterson, M. (1993) The Politics of Climate Change after UNCED. *Environmental Politics*, 2, 4, Winter 1993.

Pearce, F. (1992) Last chance to save the planet? *New Scientist*, 30 May.

Pearce, F. (1995) Dambusters. *New Internationalist*, November, 30.

Philip, G. (1982) *Oil and Politics in Latin America: Nationalist Movements and State Companies*. Cambridge: Cambridge University Press.

Philip, G. (1994) *The Political Economy of International Oil*. Edinburgh: Edinburgh University Press.

Pike, E. G. (1987) *Engineering against Schistosomiasis/Bilharzia: Guidelines towards Control of the Disease*. Basingstoke: Macmillan.

Pimentel, D. et al., (1987) World agriculture and soil erosion. *BioScience*, 37(4), 277–83.

Plant, R. (1978) *Guatemala: Unnatural Disaster*. London: Latin American Bureau.

Porter, G. and Brown, J. W. (1991) *Global Environmental Politics*. Boulder, CO: Westview.

Potter, R. B. and Salau, A. T. (eds) (1990) *Cities and Development in the Third World*. London: Mansell.

Potter, R. B. (1990) Shelter in urban Barbados, West Indies: vernacular architecture, land-tenure and self-help. In R. B. Potter and A. T. Salau (eds), *Cities and Development in the Third World*. London: Mansell.

Prance, G. (1989) Economic prospects from tropical rainforest ethnobotany. In J. O. Browder (ed.), *Fragile Lands of Latin America: Strategies for Sustainable Development*. Boulder, CO: Westview with Roger Thayer Stone Center for Latin American Studies of Tulane University.

Prins, G. (ed.) (1993) *Threats without Enemies*. London: Earthscan.

Programme for Promoting Nuclear Non-proliferation (1998) *Newsbrief*, 42, second quarter. Southampton: University of Southampton, Mountbatten Centre for International Studies.

Putman, R. D. and Bayne, N. (1987) *Hanging Together: Cooperation and Conflict in the Seven-power Summits*, rev. edn. London: Sage.

Qu, G. and Li, J. (1994) *Population and the Environment in China*. London: Paul Chapman.

Rayner, S. (1995) Governance and the global commons. In M. Desai and P. Redfern (eds), *Global Governance: Ethics and Economics of the World Order*. London: Pinter.

Reed, D. (ed.) (1992) *Structural Adjustment and the Environment*. London: Earthscan Publications.

Regens, J. L. and Rycroft, R. W. (1988) *The Acid Rain Controversy*. Pittsburgh: University of Pittsburgh Press.

Reid, D. (1995) *Sustainable Development: an Introductory Guide*. London: Earthscan.

Repetto, R. (1979) *Economic Equality and Fertility in Developing Countries*. Baltimore, MD: Johns Hopkins University Press for Resources for the Future.

Rowlands, Ian H. (1998) *The politics of global environmental change*. Manchester, Manchester University Press.

Rumaihi, M. (1986) *Beyond Oil: Unity and Development in the Gulf*. London: Al Saqi Books.

Rustow, D. A. and Mugno, J. F. (1976) *OPEC: Success and Prospects*. London: Martin Robertson.

Sachs, W. (ed.) (1992) *The Development Dictionary: a Guide to Knowledge as Power*. London: Zed Books.

Sachs, W. (ed.) (1993) *Global Ecology: a New Arena of Political Conflict*. London: Zed Books.

Salau, A. T. (1990) Urbanization and spatial strategies in West Africa in R. B. Potter and A. T. Salau (eds), *Cities and Development in the Third World*. London: Mansell.

Sampson, A. (1975) *The Seven Sisters: the Great Oil Companies and the World They Made*. London: Coronet.

Sarre, P. and Blunden, J. (eds) (1995) *An Overcrowded World? Population, Resources, and the Environment*. Oxford: Oxford University Press for the Open University.

Seabrook, J. (1996) *In the Cities of the South: Scenes from a Developing World*. London: Verso.

Sen, A. (1981) *Poverty and Famines*. Oxford: Clarendon Press.

Shiva, V. (1989) *Staying alive: women, ecology and development*. London, Zed Books.

Simeons, C. (1978) *Coal: Its Role in Tomorrow's Technology*. Oxford: Pergamon Press.

Simon, D. (1992) *Cities, Capital and Development: African Cities in the World Economy*. London: Belhaven Press.

Simon, J. (1980) Resources, population, environment: an oversupply of false bad news. *Science*, 208, 1431–7.

Smith, B. D. (1988) *State Responsibility and the Marine Environment: the Rules of Decision*. Oxford: Clarendon Press.

Spector, B. I., Sjostedt, G. and Zartman, I. W. (eds) (1994) *Negotiating International Regimes: Lessons Learned from the UNCED*.

Stoeckel, J. and Jain, A. (1986) *Fertility in Asia: Assessing the Impact of Development Projects*. London: Pinter.

Swift, R. (1995) Flood of protest. *New Internationalist*, November.

Tapp, J. F., Hunt, S. M. and Wharfe, J. R. (1996) *Toxic Impacts of Wastes on the Aquatic Environment*. London: The Royal Society of Chemistry.

Tempest, P. (ed.) (1993) *The Politics of Middle East Oil: the Royaumont Group*. London: Graham & Trotman.

Theutenberg, B. J. (1984) *The Evolution of the Law of the Sea: a Study of Resources and Strategy with Special Regard to the Polar Regions*. Dublin: Tycooly International.

Third World Resurgence 72/73.

Thomas, C. (1992) *The Environment in International Relations*. London: Royal Institute of International Affairs.

Thompson, C. P. A. (ed.) (1993) *Offshore Loss Prevention: a Systematic Approach*. London: Mechanical Engineering Publications Limited.

Tobin, R. J. (1994) Environment, population, and economic development in N. J. Vig and M. E. Kraft (eds), *Environmental Policy in the 1990s*. Washington, DC: Congressional Quarterly.

Trudgill, S. T. (1977) *Soil and Vegetation Systems*. Oxford: Clarendon Press.

Tugendhat, C. and Hamilton, A. (1975) *Oil: the Biggest Business*. London: Eyre Methuen.

Turner, L. (1978) *Oil Companies in the International System*. London: George Allen & Unwin for Royal Institute of International Affairs.

United Nations (1958) The future growth of world population. *Population Studies*, 28.

United Nations (1974) *Yearbook of the United Nations 1974*. New York: United Nations.

United Nations (1992) *Adoption of Agreements on Environment and Development: Note by the Secretary-General of the Conference*. A/CONF, 151/7, 4 June. New York: United Nations.

UN Centre for Human Settlements (Habitat) (1996) *An Urbanizing World: Global Report on Human Settlements 1996*. Nairobi UNCHS (Habitat).

UN Development Programme (1994) *Global Environment Facility: Independent Evaluation of the Pilot Phase*. Washington, DC: World Bank with UNDP, UNEP.

UNFPA (1989) *The State of World Population, 1989*. New York: United Nations Population Fund.

UNFPA (1992) *The State of World Population, 1992*. New York: United Nations Population Fund.

Uvin, P. (1996) Scaling up the grassroots and scaling down the summit: the relations between Third World NGOs and the UN in T. G. Weiss and L. Gordenker (eds), *NGOs, the UN, and Global Governance*. Boulder, CO: Lynne Reinner.

Vajpeyi, D. (1994) To dam or not to dam: socioeconomic and political impact of large hydroelectric projects – India's Narmada Basin project. Paper presented to XVI World Congress, International Political Science Association, Berlin, Germany, 21–25 August.

Vig, N. J. and Kraft, M. E. (eds) (1994) *Environmental Policy in the 1990s*. Washington, DC: Congressional Quarterly.

Vitousek, P. M., Ehrlich, P. R., Ehrlich, A. H. and Matson, P. A. (1986) Human appropriation of the products of photosynthesis. *BioScience*, 34(6), pp. 368–73.

Waller, G. (ed.) (1996) *A Complete Guide to the Marine Environment*. Mountfield, East Sussex: Pica Press.

Wang, S. and Zhao, X. (1992) Sea-level changes in China – past and future: their impact and countermeasures in G. J. H. McCall, D. J. C. Laming and S. C. Scott (eds), *Geohazards: Natural and Man-made*. London: Chapman & Hall.

Weir, D. (1988) *The Bhopal Syndrome: Pesticides, Environment and Health*. London: Earthscan.

Weiss, T. G. and Gordenker, L. (1996) *NGOs, the UN, and Global Governance*. Boulder, CO: Lynne Reinner.

Wellburn, A. (1988) *Air Pollution and Acid Rain: the Biological Impact*. London: Longman.

Wijkman, A. and Timberlake, L. (1984) *Natural Disasters: Acts of God or Acts of Man?* London: Earthscan for the International Institute for Environment and Development and the Swedish Red Cross.

Williams, M. (1994) *International Economic Organisations and the Third World*. Hemel Hempstead: Harvester Wheatsheaf.

Wood, B. (1966) *The United States and Latin American Wars, 1932–1942*. New York: Columbia University Press.

World Bank (1992) *World Development Report 1992*. New York: Oxford University Press.

World Bank (1994) *World Development Report 1994*. New York: Oxford University Press.

World Bank, Global Environment Facility (1992a) *A Bulletin on the Global Environmental Facility*, No. 5, May. Washington, DC: World Bank.

World Bank, Global Environment Facility (1992b) *A Selection of Projects from the First Three Tranches*. Working Paper Series No. 2, June. Washington, DC: GEF/UNDP.

World Commission on Environment and Development (1987) *Our Common Future (The Brundtland Report)*. Oxford: World Commission on Environment and Development/Oxford University Press.

World Health Organization (1992) *Our Planet, Our Health: Report of the WHO Commission on Health and Environment*. Geneva: WHO.

Worsley, P. (1967) *The Third World*. London: Weidenfeld & Nicholson.

Young, O. R. (1994) *International Governance*. Ithaca, NY: Cornell University Press.

INDEX

Bold entries signify principal entries